General Principles of the Method of Least Squares

Dana P. Bartlett, S. B.,

Professor of Mathematics
Massachusetts Institute of Technology

Dover Publications, Inc.
Mineola, New York

DOVER PHOENIX EDITIONS

Bibliographical Note

This Dover edition, first published in 2006, is an unabridged republication of the 1915 third edition of the work originally published by the author, Boston, Massachusetts.

International Standard Book Number: 0-486-45079-1

Manufactured in the United States of America
Dover Publications, Inc., 31 East 2nd Street, Mineola, N.Y. 11501

PREFACE.

The preparation of this volume was undertaken with the view of presenting in as simple and concise a manner as possible the fundamental principles of the Method of Least Squares. While it is believed that everything essential to the solution of all ordinary problems has been included, no attempt has been made to develop at length those special methods and forms that are so useful and almost necessary in case large numbers of observations of certain kinds, such, for instance, as those met with in geodetic and astronomical measurements, are to be adjusted.

Frequent references throughout the text, and more particularly the list of works given on page v of the Appendix, will, however, enable the student to extend his studies in whatever special direction his profession may require; it being expected that this book will in such cases be looked upon merely as an introductory treatise. All of the works mentioned have been freely consulted in the preparation of these pages, and the author desires in particular to acknowledge his indebtedness for many of the examples.

DANA P. BARTLETT.

CONTENTS.

CHAPTER I.

GENERAL PRINCIPLES.

CHAPTER II.

THE ADJUSTMENT OF OBSERVATIONS.

CHAPTER III.

THE PRECISION OF OBSERVATIONS.

CHAPTER IV.

COMPUTATION OF THE PRECISION MEASURES.

CHAPTER V.

MISCELLANEOUS THEOREMS.

CHAPTER VI.

GAUSS'S METHOD OF SUBSTITUTION.

APPENDIX.

THE THEORY OF PROBABILITY.

THE METHOD OF LEAST SQUARES.

CHAPTER I.

GENERAL PRINCIPLES.

1. In scientific investigations of all kinds it is frequently necessary to determine the values of certain quantities by means of actual measurements either with or without the aid of instruments. The observations may be made directly upon the values of the unknown quantities or upon certain functions of the unknowns. In the latter case the values of the required quantities must be obtained by computation from the observed values of the functions. In order to obtain more accurate values of the unknowns than would be given by a single measurement, or set of measurements, the observations are usually repeated either in the same way and under the same conditions or in a variety of different ways and under varying conditions.

Under these circumstances it will invariably be found that the different measurements give discordant results, the amount of the discrepancies varying with the character of the observations; and the question that now presents itself is how to determine from these discordant observations the true values of the required quantities. From the nature of the case, however, we can not expect to obtain our values with absolute accuracy; all that we can hope for is to obtain those values which are rendered most probable after all the observations are taken into account, and, further, to determine the degree of confidence that can be placed in those values.

2. The attainment of the above results constitutes the primary object of the Method of Least Squares. The method is also employed in comparing the relative worth of different measurements of the same quantity, and in determining the equation of a curve which shall suitably represent the relation between two variables in cases where the exact law connecting them is not known.

Also, before making any observations, we may employ the method to determine how precise the component measurements of a series must be in order to yield a required degree of precision in the final result; or, conversely, to determine what the precision of the final result will be, knowing the precision attainable in the component measurements. This latter application of the method will be treated at length in the course on "The Precision of Measurements."

3. Errors. The cause of the discrepancies between the results of our different observations is that every observation that is a measure is subject to error. These errors are of two kinds,—Constant or Systematic Errors and Accidental Errors.

4. Constant Errors are errors which in all measures of the same quantity, made with the same care and under the same conditions, have the same magnitude, or whose presence and magnitude are due to some fixed cause. These constant errors may be of several classes, which are designated as follows:—

First. **Theoretical Errors,** such as those due to the refraction or aberration of light, the effect of a definite change in temperature or moisture on our standards of measurement, etc. As soon as their causes are known the magnitude of these errors may be calculated and their effect eliminated from the observations.

Second. **Instrumental Errors,** such as errors of division of graduated scales, defects in micrometer screws, eccentricity of circles, etc. These errors will be discovered by an examination of the instruments and their effects eliminated from the observations, either by a particular method of using the instruments or by subsequent computation.

Third. **Personal Errors.** These are due to personal peculiarities of an observer, who always answers a signal too soon or too late, always estimates a quantity smaller than it is, etc. The character and magnitude of these errors may be determined by a study of the observer, his "Personal Equation" may be obtained, and his observations thus corrected for this source of error.

5. **Mistakes.** Although of a somewhat different character, these should be considered in connection with constant errors. A mistake is made when a figure 3 is read for a figure 8, or when in reading a graduated circle which is numbered in both directions the angle is read 43° instead of the complementary angle 47°, etc. These mistakes are usually of such a character that they may be detected by an inspection of the observations and a proper correction made.

6. **Accidental Errors** are errors due to irregular causes, whose effect upon the observations is not determined by any circumstances peculiar to that particular set of measurements, and which cannot therefore be computed and allowed for beforehand. Such errors are those due to sudden changes in refraction owing to sudden and unobserved changes in temperature; unequal expansion of different parts of an instrument with change in temperature; shaking of an instrument in the wind, etc. But most important of all are those errors which arise from imperfections in the sight, hearing, and other senses of the observer, which render it impossible for him to adjust and use his instruments with absolute accuracy.

After a full investigation of the constant errors, the observer should diminish the accidental errors as much as possible, both in number and magnitude, by taking every precaution and care in the measurements themselves. The problem now remains to combine the observations so that the remaining accidental errors shall have the least probable effect upon the results, and it is to bring about this combination of observations that we employ the Method of Least Squares.

When no more observations are made than are sufficient to determine one value for each of the unknown quantities, we must accept these values as the most probable ones. But if additional observations are made leading to discordant results, we can not take any one of them as the correct value, and in fact, as already stated, we shall probably not be able to obtain the true values of the unknowns. All that we can do is to find values of the unknowns which shall remove the discrepancies between the different observations and which shall be those values that are rendered most probable by the existence of the observations themselves.

On first thoughts it may seem that these accidental errors, being due to so many different and unknown causes, will be beyond the scope of mathematical investigation. Nevertheless, the theory of probability requires that these errors shall follow in magnitude and frequency a law that is capable of exact mathematical expression, and experience confirms the correctness of this law.

For more extended remarks on these subjects see —

Holman, "Discussion of the Precision of Measurements," pp. 1–14.
Merriman, "Text-Book of Least Squares," pp. 1–6.
Chauvenet, "Spherical and Practical Astronomy," pp. 469–473.
Wright, "Treatise on the Adjustment of Observations," pp. 11–18.

LAWS OF ERRORS OF OBSERVATION.

7. The derivation of the general laws of the occurrence of errors of observation, and of the processes for determining the most probable values of the unknown quantities, will be based upon the following

Axiom. If a series of n direct observations, $M_1, M_2, \dots M_n$, is made upon the value of a quantity M, all the observations being made with the same care and under the same circumstances, the most probable value M_0 of that quantity is the arithmetical mean of the observations. Or

$$M_0 = \frac{M_1 + M_2 + \dots M_n}{n} = \frac{\Sigma M}{n} \qquad (1)$$

8. The **Real Error** (x) of an observation is the difference between the observed value of the measured quantity and the real value.

9. The **Residual** (v) of an observation is the difference between the observed value of the measured quantity and the value rendered most probable by the existence of the observations.

10. *Example.* Eight observations are made upon the resistance of a coil of wire, the true resistance being 512. Find from these observations the most probable resistance, and also the real errors and residuals.

	Observations. $M.$	Real Errors. x		Residuals. v
	512.4	+.4		+.30
	512.2	+.2		+.10
	511.9	−.1		−.20
	512.3	+.3		+.20
	511.8	−.2		−.30
	512.3	+.3		+.20
	511.9	−.1		−.20
	512.0	.0		−.10
Mean =	512.10		$\Sigma v =$	.00

From the observations, then, we should say that the most probable resistance of the coil is 512.10. It will also be noticed that the sum of the residuals is zero. That this is a general result following from the assumption of the arithmetical mean as the most probable value may be proved as follows: If the observations are M_1, M_2, . . . M_n, the arithmetical mean M_0, and the residuals $v_1, v_2, \ldots v_n$, then we have

$$v_1 = M_1 - M_0, \quad v_2 = M_2 - M_0, \quad \ldots v_n = M_n - M_0.$$

$$\Sigma v = \Sigma M - nM_0$$

$$= \Sigma M - \Sigma M, \quad \text{since } M_0 = \frac{\Sigma M}{n}$$

$$\therefore \quad \Sigma v = 0 \qquad (2)$$

11. Weighted Observations. The weight of an observation expresses its relative worth compared with other observations. Thus, if six observations are made upon the value of a quantity, five of which give the same result, while the sixth differs, in combining these two different results to obtain the most probable value of the unknown, the first value ought to have five times the influence upon the final result that the second has, since it has taken five times as much labor and time to obtain it. Hence in general we may say —

12. The **Weight** (p) of an observation may be considered as representing the number of times the observation has been repeated and the same result obtained.

The weights assigned to observations may be due to a variety of causes, as difference in skill of observers, difference in the instruments used or the circumstances under which the observations are made, etc. But whatever the cause, the *effect* on the final values of weighting an observation will be the same as indicated in the preceding paragraph.

13. *Example.* Suppose n observations, $M_1, M_2, \ldots M_n$, of weights $p_1, p_2, \ldots p_n$, are made upon the value of a quantity M. To find the most probable value M_0 of the quantity.

From the above interpretation of the meaning of weight, we may consider that the whole number of observations is $p_1 + p_2 + \ldots p_n$, or Σp, and that the result M_1 has been obtained in p_1 observations, M_2 in p_2 observations, etc., Therefore, by (1)

$$M_0 = \frac{p_1 M_1 + p_2 M_2 + \ldots p_n M_n}{\Sigma p} = \frac{\Sigma p M}{\Sigma p} \qquad (3)$$

M_0 is called the *General Mean.*

If the residuals are $v_1, v_2, \ldots v_n$, we have

$$v_1 = M_1 - M_0, \quad v_2 = M_2 - M_0, \ldots v_n = M_n - M_0,$$

$$p_1 v_1 + p_2 v_2 + \ldots p_n v_n = p_1 M_1 + p_2 M_2 + \ldots p_n M_n - M_0 \Sigma p$$

$$\therefore \quad \Sigma p v = 0 \qquad (4)$$

Which shows that in the case of direct observations of different weights the sum of the weighted residuals is zero.

14. If the observations are not made directly upon the values of the required quantities, the method of adjusting the results so as to obtain the best possible values of the unknowns will depend upon the laws which govern the distribution of the errors of these observations. It is found in practice that the accidental errors of observations follow certain well defined laws, and what these are may best be seen by taking an actual example.

15. *Example.* One thousand shots are fired at a target which is divided into a number of horizontal sections by lines one foot apart, the centre line of the target being in the middle of one of these spaces. The shots were distributed as follows:—

In Space.	*Shots.*	*In Space.*	*Shots.*	*In Space.*	*Shots.*
$+5\frac{1}{2}$ to $+4\frac{1}{2}$	1	$+1\frac{1}{2}$ to $+\frac{1}{2}$	190	$-2\frac{1}{2}$ to $-3\frac{1}{2}$	79
$+4\frac{1}{2}$ " $+3\frac{1}{2}$	4	$+\frac{1}{2}$ " $-\frac{1}{2}$	212	$-3\frac{1}{2}$ " $-4\frac{1}{2}$	16
$+3\frac{1}{2}$ " $+2\frac{1}{2}$	10	$-\frac{1}{2}$ " $-1\frac{1}{2}$	204	$-4\frac{1}{2}$ " $-5\frac{1}{2}$	2
$+2\frac{1}{2}$ " $+1\frac{1}{2}$	89	$-1\frac{1}{2}$ " $-2\frac{1}{2}$	193		

In this case the errors are evidently the distances of the shots from the centre of the target. Further, as far as can be judged from these one thousand shots, if another shot is fired the probability that this shot will fall between the lines

$+5\frac{1}{2}$ and $+4\frac{1}{2}$ is .001		$-\frac{1}{2}$ and $-1\frac{1}{2}$ is .204
$+4\frac{1}{2}$ " $+3\frac{1}{2}$ " .004		$-1\frac{1}{2}$ " $-2\frac{1}{2}$ " .193
$+3\frac{1}{2}$ " $+2\frac{1}{2}$ " .010		$-2\frac{1}{2}$ " $-3\frac{1}{2}$ " .079
$+2\frac{1}{2}$ " $+1\frac{1}{2}$ " .089		$-3\frac{1}{2}$ " $-4\frac{1}{2}$ " .016
$+1\frac{1}{2}$ " $+\frac{1}{2}$ " .190		$-4\frac{1}{2}$ " $-5\frac{1}{2}$ " .002
$+\frac{1}{2}$ " $-\frac{1}{2}$ " .212		

The sum of the above probabilities is unity, and, therefore, as far as the preceding shots show the 1001st shot will certainly hit the target.

16. Now using as abscissas the distances of the horizontal lines from the centre of the target, and as ordinates the number of shots falling in the corresponding spaces, we may construct the following figure:—

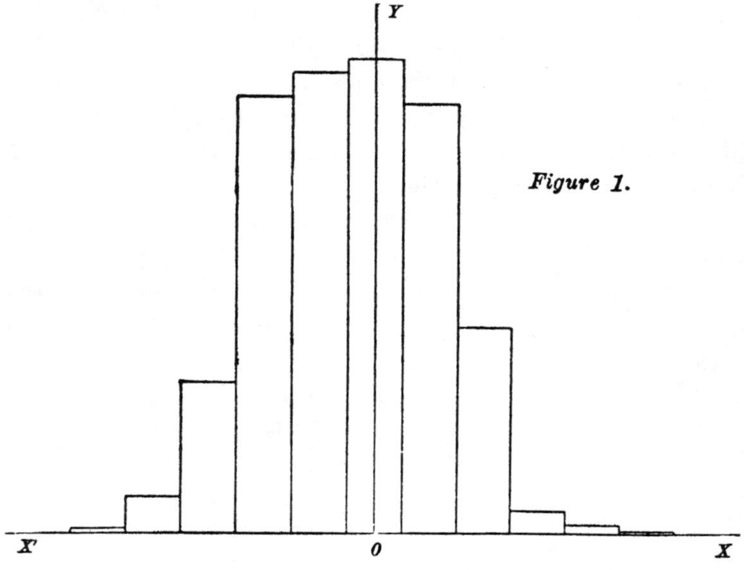

Figure 1.

And if the entire area of this figure is taken as unity then the area of each rectangle will denote the probability of a

shot, if fired, falling within the corresponding space of the target.

The graphical representation of the accidental errors of observation will always give a figure similar to the above. Hence denoting errors by abscissas, and their frequency by ordinates, the law of error of any series of observations may be represented by a curve whose general form is determined by Figure 1. This curve is called the "Curve of Error," and is shown in Figure 2.

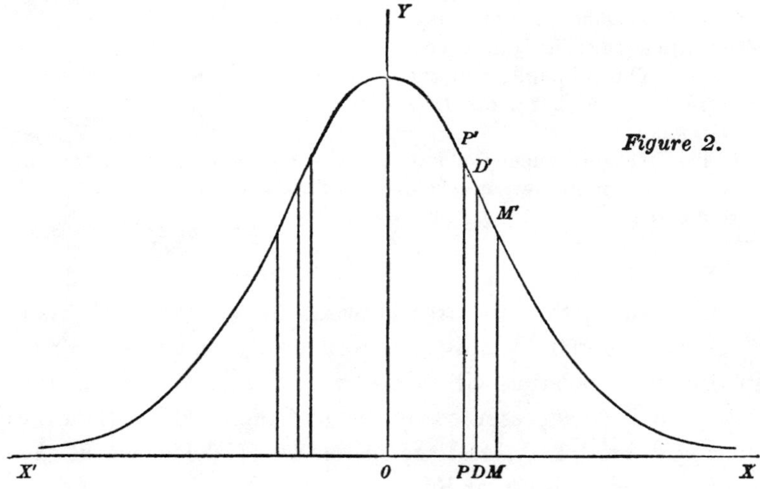

Figure 2.

In order that this curve may represent exactly the distribution of the errors in any given series of observations it ought to meet the axis of X at some definite distance to the right and left of the origin and coincide with the axis from there on, for in all actual observations there is a limit beyond which no errors occur. But as the exact point of meeting could not be determined for any given case, and as it would not be possible to obtain the equation of such a curve, we make it asymptotic to the axis of X, taking care that the error thus introduced shall in any set of observations be so small as to be negligible.

17. An inspection of Figures 1 and 2 will now exhibit some of the general laws of errors of observation and the corresponding properties of the curve of error.

Laws of Error derived from an inspection of Figure 1.	*Representation of these laws by the Curve of Error.*
First. Small errors are more frequent than large ones.	The maximum point of the curve is on the axis of Y.
Second. Positive and negative errors of the same absolute magnitude are equally likely to occur.	The curve is symmetrical with respect to the axis of Y.
Third. The probability of the occurrence of very large errors is very small.	The curve is asymptotic to the axis of X.
Fourth. The frequency of any error depends upon the magnitude of that error.	The equation of the curve will be of the form $$y = \phi(x) \qquad (5)$$

18. If, now, the total area between the curve and the axis of X be denoted by unity the probability that the error of any given observation will fall between the magnitudes x and $x + dx$ will be represented by the area included between the curve, the axis of X, and the ordinates of the curve at the errors x and $x + dx$, or by

$$y\ dx = \phi(x)\ dx \qquad (6)$$

And this probability will be known as soon as we find the form of the function $\phi(x)$.

19. The above expressions in (5) and (6) are the ones that we should use if we regard the curve of error as representing the law of occurrence of errors of observation. If, however, we look upon the curve as expressing the law to which we must make the residuals conform, in order that the values of the unknown quantities obtained from them may be the most probable values, we should replace x by v and use the expressions

$$y = \phi(v) \tag{7}$$

and

$$y\,dv = \phi(v)\,dv \tag{8}$$

for (5) and (6), respectively.

THE EQUATION OF THE CURVE OF ERROR.

20. Let n observations, all of the same weight, with results $M_1, M_2, \ldots M_n$, be made upon any function or functions of a number of unknown quantities $z_1, z_2, \ldots z_q$; and let the residuals of $M_1, M_2, \ldots M_n$ be $v_1, v_2, \ldots v_n$, and the probability of the occurrence of these residuals be $\phi(v_1)\,dv$, $\phi(v_2)\,dv$, $\ldots\ \phi(v_n)\,dv$, respectively. Then the probability of the simultaneous occurrence of all these residuals will be

$$P = \phi(v_1)\,\phi(v_2)\ldots\phi(v_n)\,(dv)^n \tag{9}$$

$$\therefore\ \log P = \log\phi(v_1) + \ldots \log\phi(v_n) + n\log dv \tag{10}$$

Each different method that might be adopted for computing the values of the unknowns $z_1, z_2, \ldots z_q$ would lead to a different set of residuals $v_1, v_2, \ldots v_n$; but obviously that set of values of $z_1, z_2, \ldots z_q$ should be considered the best which corresponds to the particular set of residuals $v_1, v_2, \ldots v_n$, the probability of whose occurrence is greater than that of any other set.

Therefore the most probable values of $z_1, z_2, \ldots z_q$ are those that make P in (9), or $\log P$ in (10), a maximum.

The values of $z_1, z_2, \ldots z_q$ corresponding with this latter condition are those that satisfy equations (11). It may be noticed that these equations also express the preliminary conditions leading to a minimum value of $\log P$, but the nature of the problem is such that a maximum value of P evidently exists while a minimum does not, and it is therefore unnecessary to investigate further the mathematical conditions for a maximum. Hence we have

$$
\left.
\begin{aligned}
\frac{\partial \log P}{\partial z_1} &= \frac{1}{\phi(v_1)}\frac{\partial \phi(v_1)}{\partial z_1} + \cdots \frac{1}{\phi(v_n)}\frac{\partial \phi(v_n)}{\partial z_1} = 0 \\
\frac{\partial \log P}{\partial z_2} &= \frac{1}{\phi(v_1)}\frac{\partial \phi(v_1)}{\partial z_2} + \cdots \frac{1}{\phi(v_n)}\frac{\partial \phi(v_n)}{\partial z_2} = 0 \\
&\cdots \qquad \cdots \qquad \cdots \\
&\cdots \qquad \cdots \qquad \cdots \\
\frac{\partial \log P}{\partial z_q} &= \frac{1}{\phi(v_1)}\frac{\partial \phi(v_1)}{\partial z_q} + \cdots \frac{1}{\phi(v_n)}\frac{\partial \phi(v_n)}{\partial z_q} = 0
\end{aligned}
\right\} \qquad (11)
$$

But

$$\frac{\partial \phi(v)}{\partial z} = \phi'(v)\frac{\partial v}{\partial z} \qquad (12)$$

and for convenience we may put

$$\frac{\phi'(v)}{\phi(v)} = \psi(v) \qquad (13)$$

substituting in (11) we have

$$
\left.
\begin{aligned}
\psi(v_1)\frac{\partial v_1}{\partial z_1} + \psi(v_2)\frac{\partial v_2}{\partial z_1} + \cdots \psi(v_n)\frac{\partial v_n}{\partial z_1} = 0 \\
\psi(v_1)\frac{\partial v_1}{\partial z_2} + \psi(v_2)\frac{\partial v_2}{\partial z_2} + \cdots \psi(v_n)\frac{\partial v_n}{\partial z_2} = 0 \\
\cdots \qquad \cdots \qquad \cdots \\
\cdots \qquad \cdots \qquad \cdots \\
\psi(v_1)\frac{\partial v_1}{\partial z_q} + \psi(v_2)\frac{\partial v_2}{\partial z_q} + \cdots \psi(v_n)\frac{\partial v_n}{\partial z_q} = 0
\end{aligned}
\right\} \qquad (14)
$$

These equations contain all the unknowns $z_1, z_2, \ldots z_q$, and there are as many equations as unknowns, hence as soon as we find the form of the function $\psi(v)$ we can solve these equations for the most probable values of $z_1, z_2, \ldots z_q$. Since we have considered the general case, and the above results are to hold true whatever the number of unknown quantities and the form of the functions observed, we may deduce the form of $\psi(v)$ by solving a special example.

Example. Let n observations of equal weight be made upon the value of a single unknown z_1, with results M_1, M_2, $\ldots M_n$, and let the residuals be $v_1, v_2, \ldots v_n$. Then the most probable value of z_1 is given by

$$z_1 = M_1 - v_1 = M_2 - v_2 = \ldots M_n - v_n$$

differentiating with respect to z_1,

$$1 = -\frac{\partial v_1}{\partial z_1} = -\frac{\partial v_2}{\partial z_1} = \ldots -\frac{\partial v_n}{\partial z_1} \qquad \text{(a)}$$

substituting (a) in (14), changing all the signs, we have

$$\psi(v_1) + \psi(v_2) + \ldots \psi(v_n) = 0 \qquad \text{(b)}$$

But in this case, as was shown in (2),

$$v_1 + v_2 + \ldots v_n = 0 \qquad \text{(c)}$$

In order that (b) and (c) may both be true the functional symbol ψ must indicate multiplication by a constant. That is, in general

$$\psi(v) = cv \qquad (15)$$

Substituting this in (13) and (12),

$$\frac{\partial \phi(v)}{\partial z} = \phi(v)\, cv\, \frac{\partial v}{\partial z}$$

therefore $$\frac{1}{\phi(v)} \frac{\partial \phi(v)}{\partial z} = cv \frac{\partial v}{\partial z}$$

Integrating, $$\log \phi(v) = \tfrac{1}{2}cv^2 + c'$$

$$\therefore \quad \phi(v) = e^{\frac{1}{2}cv^2 + c'}$$

$$= ke^{\frac{1}{2}cv^2}$$

Since $y = \phi(v)$ is the equation of the curve of error, (7), we may therefore write it

$$y = ke^{\frac{1}{2}cv^2}$$

But on examination of the curve, y is seen to be a decreasing function of v, and hence the exponent of e is essentially negative. Accordingly we will write our equation in the form

$$y = ke^{-h^2v^2} \qquad (16)$$

the values of k and h depending upon the character of the observations, but in all observations of the same kind and weight having the same values.

This equation represents the law in accordance with which the residuals must be distributed in order that the best results may be obtained from our observations. But, as before mentioned, if we wish our curve to represent the most probable distribution of the real errors of observation we should write the equation in the form

$$y = ke^{-h^2x^2} \qquad (17)$$

Hereafter we shall use without further remark either form of the equation according to the aspect in which we are considering our curve.

An inspection of the above equation will show that it satisfies all the conditions noted in discussing the form of the curve of error in paragraph 17.

21. It is important to notice that in all discussions in the Method of Least Squares the number of observations is supposed to be large and always greater than the number of unknown quantities. As will be illustrated later on, paragraph 91, whenever this is the case there is a remarkable agreement between the results obtained in practice and those indicated by the theory. And even when the observations are few in number the method still affords the best means at our command for their adjustment, the results obtained merely having a smaller weight than they would have had if derived from a greater number of observations.

THE METHOD OF LEAST SQUARES.

22. We are now in a position to see whence comes the name "Least Squares."

In paragraph 20 it was pointed out that whenever we make a series of observations, each observation of the set having the same weight, the most probable system of values of the unknown quantities will be that which corresponds with the set of residuals the probability of whose occurrence is a maximum. That is, the best set of values of the unknowns will be that which gives a maximum value to

$$P = \phi(v_1)\ \phi(v_2)\ \ldots\ \phi(v_n)\ (dv)^n$$

But from equation (16) this reduces to

$$P = k^n e^{-h^2(v_1^2 + v_2^2 + \ldots v_n^2)}\ (dv)^n \qquad (18)$$

Since the exponent of e in this expression is negative, evidently P will be a maximum when

$$v_1^2 + v_2^2 + \ldots v_n^2 = \Sigma v^2 \textit{ is a minimum.} \qquad (19)$$

Hence the adjustment of observations by the Method of Least Squares is based upon the principle that the most probable system of values of the unknowns is that which renders the sum of the squares of the residuals a minimum. Hence the name.

The conditions for a maximum value of P were expressed in equations (14), and since it has been shown that the function ψ means multiplication by a constant, those equations reduce to the following, called

NORMAL EQUATIONS.

$$
\left.
\begin{array}{l}
v_1 \dfrac{\partial v_1}{\partial z_1} + v_2 \dfrac{\partial v_2}{\partial z_1} + \ldots v_n \dfrac{\partial v_n}{\partial z_1} = 0 \\
v_1 \dfrac{\partial v_1}{\partial z_2} + v_2 \dfrac{\partial v_2}{\partial z_2} + \ldots v_n \dfrac{\partial v_n}{\partial z_2} = 0 \\
\quad \ldots \qquad \ldots \qquad \ldots \\
\quad \ldots \qquad \ldots \qquad \ldots \\
v_1 \dfrac{\partial v_1}{\partial z_q} + v_2 \dfrac{\partial v_2}{\partial z_q} + \ldots v_n \dfrac{\partial v_n}{\partial z_q} = 0
\end{array}
\right\} \qquad (20)
$$

An inspection of these equations will show that they also express the conditions that will make the sum of the squares of the residuals a minimum.

In the adjustment of observations the above are the fundamental equations. In order to obtain the most probable values of the unknowns in any set of observations, all that is necessary is to form the Normal Equations for that set and solve them simultaneously. The examples already solved for direct observations are merely special cases of the above general solution.

CHAPTER II.

THE ADJUSTMENT OF OBSERVATIONS.

INDIRECT OBSERVATIONS.

23. In the determination of the values of quantities by means of observations the functions of the unknowns that it is necessary to observe may be of any form, but if they are not linear the normal equations derived from them are likely to be complicated and difficult, if not impossible, to solve. Hence if the observations are not upon linear functions of the unknowns, the first step will be to reduce them to equivalent linear expressions by transformations depending on the character of the observed functions; see paragraphs 43 and 44. It will be necessary to consider, therefore, the method of adjusting observations on linear functions alone, and the procedure in cases of this kind may be illustrated by the following simple example.

24. *Example.* S_1, S_2, S_3, are three solids whose masses are required. Not having standard weights enough to obtain all these masses directly, by varying the distribution of the solids in the pans of the balance the following observations are made:—

$$\begin{aligned} S_1 &= S_2 + 1.7 \text{ grams.} \\ S_3 &= \phantom{S_2 +{}} 2.4 \quad \text{"} \\ S_2 + S_3 &= S_1 + 1.0 \quad \text{"} \\ S_2 &= S_3 + 3.0 \quad \text{"} \end{aligned}$$

If, now, the most probable values of S_1, S_2, S_3, are represented by z_1, z_2, z_3, and the corresponding residuals of the observations by v_1, v_2, . . . v_4, the values of these residuals in terms of z_1, z_2, z_3 may be found from the above observations by transposing all terms to the first members of the equations, and we obtain at once the following, called

OBSERVATION EQUATIONS.

$$\begin{aligned} z_1 - z_2 \qquad\quad - 1.7 &= v_1 \\ z_3 - 2.4 &= v_2 \\ -z_1 + z_2 + z_3 - 1.0 &= v_3 \\ z_2 - z_3 - 3.0 &= v_4 \end{aligned} \qquad \text{(A)}$$

Applying equations (20) as formulas we have the

NORMAL EQUATIONS.

$$\begin{aligned} &(z_1 - z_2 - 1.7) + (-z_1 + z_2 + z_3 - 1.0)(-1) = 0 \\ &(z_1 - z_2 - 1.7)(-1) + (-z_1 + z_2 + z_3 - 1.0) \\ &\qquad\qquad + (z_2 - z_3 - 3.0) = 0 \qquad \text{(B)} \\ &(z_3 - 2.4) + (-z_1 + z_2 + z_3 - 1.0) \\ &\qquad\qquad + (z_2 - z_3 - 3.0)(-1) = 0 \end{aligned}$$

Simplified, these become

$$\begin{aligned} 2z_1 - 2z_2 - z_3 - 0.7 &= 0 && \text{(a)} \\ -2z_1 + 3z_2 \qquad - 2.3 &= 0 && \text{(b)} \\ -z_1 \qquad + 3z_3 - 0.4 &= 0 && \text{(c)} \end{aligned}$$

$$\begin{aligned} 3 \times \text{(a)} \qquad 6z_1 - 6z_2 - 3z_3 - 2.1 &= 0 \\ \text{(c)} \qquad -z_1 \qquad + 3z_3 - 0.4 &= 0 \\ \hline 5z_1 - 6z_2 \qquad - 2.5 &= 0 \\ 2 \times \text{(b)} \qquad -4z_1 + 6z_2 \qquad - 4.6 &= 0 \\ \hline z_1 \qquad - 7.1 &= 0 \end{aligned}$$

$$z_1 = 7.1 \qquad \text{(d)}$$

substitute (d) in (b) $\qquad z_2 = 5.5 \qquad$ (e)

" $\quad$ (d) in (c) $\qquad z_3 = 2.5 \qquad$ (f)

An inspection of the work in this example will show that for the adjustment of observations of equal weight on linear functions of the unknowns we may derive the following:—

25. Rule. *For each observation write an "Observation Equation"; then for each unknown form a "Normal Equation," by multiplying the first member of each observation equation by the coefficient of that unknown in that equation, adding the results and placing the sum equal to zero. Solve these equations simultaneously for the values of the unknowns.*

In solving for the most probable values of the unknowns the second members of the observation equations are very commonly written zero instead of $v_1, v_2, \ldots v_n$. For this is the form in which the equations naturally appear, and if the observations were exact the residuals would actually all be zero. The method of solution is the same in either case.

26. Observations of Unequal Weight. If the observations are not all of equal weight the same method will apply, except that in the formation of the normal equations each observation equation will be used the number of times denoted by its weight. Thus in the last example if the observations have the weights 4, 9, 1, 4, the normal equations will have the same form as in (B), page 18, but each part of each equation will be multiplied by the weight of the observation equation from which it is derived. This will give the

NORMAL EQUATIONS.

$$4(z_1 - z_2 - 1.7) + (-z_1 + z_2 + z_3 - 1.0)(-1) = 0$$
$$4(z_1 - z_2 - 1.7)(-1) + (-z_1 + z_2 + z_3 - 1.0) + 4(z_2 - z_3 - 3.0) = 0$$
$$9(z_3 - 2.4) + (-z_1 + z_2 + z_3 - 1.0) + 4(z_2 - z_3 - 3.0)(-1) = 0$$

or reduced

$$5z_1 - 5z_2 - z_3 - 5.8 = 0 \quad \text{(a)}$$
$$-5z_1 + 9z_2 - 3z_3 - 6.2 = 0 \quad \text{(b)}$$
$$-z_1 - 3z_2 + 14z_3 - 10.6 = 0 \quad \text{(c)}$$

the solution of which gives

$$z_1 = 7.07 \qquad z_2 = 5.42 \qquad z_3 = 2.42$$

Further it will at once be seen that if $p_1, p_2, \ldots p_n$, are the weights of the corresponding observations, equations (20) take the general form:—

WEIGHTED NORMAL EQUATIONS.

$$
\begin{aligned}
&p_1 v_1 \frac{\partial v_1}{\partial z_1} + p_2 v_2 \frac{\partial v_2}{\partial z_1} + \ldots p_n v_n \frac{\partial v_n}{\partial z_1} = 0 \\
&p_1 v_1 \frac{\partial v_1}{\partial z_2} + p_2 v_2 \frac{\partial v_2}{\partial z_2} + \ldots p_n v_n \frac{\partial v_n}{\partial z_2} = 0 \\
&\quad\ldots \qquad \ldots \qquad \ldots \\
&\quad\ldots \qquad \ldots \qquad \ldots \\
&p_1 v_1 \frac{\partial v_1}{\partial z_q} + p_2 v_2 \frac{\partial v_2}{\partial z_q} + \ldots p_n v_n \frac{\partial v_n}{\partial z_q} = 0
\end{aligned} \qquad (21)
$$

Hence for the formation of the normal equations in weighted observations on linear functions of the unknowns, we have the following:—

27. Rule. *For each observation write an "Observation Equation"; then for each unknown form a "Normal Equation," by multiplying the first member of each observation equation by the coefficient of that unknown in that equation and by the weight of that equation, adding the results and placing the sum equal to zero. Solve these equations simultaneously.*

28. The same result will be obtained if we begin by multiplying each observation equation by the square root of its weight and then proceed according to the first rule (paragraph 25).

This result illustrates the important principle that *multiplying a set of equations by the square roots of their weights reduces them all to equivalent equations of weight unity.*

29. Relation between the Weight of an Observation and the Value of *h*. If in paragraph 20 the n observations have weights $p_1, p_2, \ldots p_n$, and the quantity h values $h_1, h_2, \ldots h_n$, then equation (18) becomes

$$P = k_1 e^{-h_1^2 v_1^2} k_2 e^{-h_2^2 v_2^2} \ldots k_n e^{-h_n^2 v_n^2} (dv)^n$$
$$= k_1 k_2 \ldots k_n e^{-(h_1^2 v_1^2 + h_2^2 v_2^2 + \ldots h_n^2 v_n^2)} (dv)^n \qquad (22)$$

The most probable set of values of the unknowns is that which makes P a maximum, and P is a maximum when

$$h_1^2 v_1^2 + h_2^2 v_2^2 + \ldots h_n^2 v_n^2 \textit{ is a minimum.} \qquad (23)$$

The conditions for a minimum value of this expression are the following, which are then for this case the

NORMAL EQUATIONS.

$$\begin{aligned} h_1^2 v_1 \frac{\partial v_1}{\partial z_1} + h_2^2 v_2 \frac{\partial v_2}{\partial z_1} + \ldots h_n^2 v_n \frac{\partial v_n}{\partial z_1} &= 0 \\ h_1^2 v_1 \frac{\partial v_1}{\partial z_2} + h_2^2 v_2 \frac{\partial v_2}{\partial z_2} + \ldots h_n^2 v_n \frac{\partial v_n}{\partial z_2} &= 0 \\ \cdots \qquad \cdots \qquad \cdots & \\ \cdots \qquad \cdots \qquad \cdots & \\ h_1^2 v_1 \frac{\partial v_1}{\partial z_q} + h_2^2 v_2 \frac{\partial v_2}{\partial z_q} + \ldots h_n^2 v_n \frac{\partial v_n}{\partial z_q} &= 0 \end{aligned} \qquad (24)$$

But equations (21) are also the normal equations for this case. Hence (21) and (24) must be identical, and

$$p_1 : p_2 : \ldots p_n = h_1^2 : h_2^2 : \ldots h_n^2 \qquad (25)$$

That is, the square of h is proportional to the weight of the observation. Accordingly, since h increases in value as the quality of the observations is improved, it is called "The Measure of Precision."

Further, it follows from (23) and (25) that the most probable system of values of the unknowns will be that in which

$$p_1 v_1^2 + p_2 v_2^2 + \ldots p_n v_n^2 \text{ is a minimum.} \qquad (26)$$

And this is the most general form of statement of the principle of "Least Squares." The same principle is represented in equations (21).

30. Computation of Corrections. If large numbers occur in the observations it is better to compute the most probable corrections to apply to the observed values rather than the most probable values of the unknowns themselves. In this way we can often avoid a large amount of numerical work.

31. *Example.* P_1, P_2, P_3, P_4, P_5, are five points whose altitudes above the mean level of the sea are to be determined from the following observations of difference of level.

$$\begin{aligned} P_1 &= 573.08 & P_4 - P_2 &= 170.28 \\ P_2 - P_1 &= 2.60 & P_4 - P_5 &= 425.00 \\ P_2 &= 575.27 & P_5 &= 319.91 \\ P_3 - P_2 &= 167.33 & P_5 &= 319.75 \\ P_4 - P_3 &= 3.80 & & \end{aligned}$$

An inspection of these observations shows that we may put

$$\begin{aligned} P_1 &= 573 + z_1 & P_4 &= 745 + z_4 \\ P_2 &= 575 + z_2 & P_5 &= 320 + z_5 \qquad (A) \\ P_3 &= 742 + z_3 & & \end{aligned}$$

where z_1, z_2, z_3, z_4, z_5, are small corrections whose most probable values are to be determined. We now have for

OBSERVATION EQUATIONS.

$$\begin{aligned} 573 + z_1 - 573.08 &= 0 \quad \text{or} \quad & z_1 - .08 &= 0 \\ 575 + z_2 - 573 - z_1 - 2.60 &= 0 \quad \text{or} \quad & z_2 - z_1 - .60 &= 0 \\ 575 + z_2 - 575.27 &= 0 \quad \text{or} \quad & z_2 - .27 &= 0 \end{aligned}$$

$$742 + z_3 - 575 - z_2 - 167.33 = 0 \quad \text{or} \quad z_3 - z_2 - .33 = 0$$
$$745 + z_4 - 742 - z_3 - 3.80 = 0 \quad \text{or} \quad z_4 - z_3 - .80 = 0$$
$$745 + z_4 - 575 - z_2 - 170.28 = 0 \quad \text{or} \quad z_4 - z_2 - .28 = 0$$
$$745 + z_4 - 320 - z_5 - 425.00 = 0 \quad \text{or} \quad z_4 - z_5 = 0$$
$$320 + z_5 - 319.91 = 0 \quad \text{or} \quad z_5 + .09 = 0$$
$$320 + z_5 - 319.75 = 0 \quad \text{or} \quad z_5 + .25 = 0$$

From these we now form the

NORMAL EQUATIONS.

$$\begin{aligned}
2z_1 - z_2 \qquad\qquad\qquad\qquad & + .52 = 0 \\
-z_1 + 4z_2 - z_3 - z_4 \qquad\quad & - .26 = 0 \\
-z_2 + 2z_3 - z_4 \qquad\quad & + .47 = 0 \\
-z_2 - z_3 + 3z_4 - z_5 & - 1.08 = 0 \\
-z_4 + 3z_5 & + .34 = 0
\end{aligned}$$

and solving,

$$z_1 = -.19; \quad z_2 = .14; \quad z_3 = .05; \quad z_4 = .43; \quad z_5 = .03$$

Substituting these in equations (A) we have for the most probable altitudes,

$$P_1 = 572.81 \qquad P_3 = 742.05 \qquad P_5 = 320.03$$
$$P_2 = 575.14 \qquad P_4 = 745.43$$

If the original observation equations had been retained, the independent terms in the normal equations would have been

570.48 240.26 163.53 599.08 214.66

32. Significant Figures. The adjustment by the Method of Least Squares of observations which occur in practice, although not difficult, is apt to be long and laborious. Hence to reduce this labor as much as possible it is of great importance that careful attention should be given in the solutions to the proper use of significant figures. When in doubt, however, as to the proper number of figures to retain it is better to keep too many rather than too few, as the superfluous figures can be rejected at the end of the computation; while if too few are retained the results obtained from the computations will be worthless.

For a general discussion of the subject of significant figures see Holman's "Precision of Measurements," pages 76 to 84, but for the present the following rules will suffice for most cases.

Rule 1. *In casting off places of figures increase by 1 the last figure retained, when the following figure is 5 or over.*

Rule 2. *In the precision measure retain two significant figures.*

Rule 3. *In any quantity retain enough significant figures to include the place in which the second significant figure of its precision measure occurs.*

Rule 4. *When several quantities are to be added or subtracted, apply Rule 3 to the least precise and keep only the corresponding figures in the other quantities.*

Rule 5. *When several quantities are to be multiplied or divided into each other, find the percentage precision of the least precise. If this is*

1 per cent or more, use four significant figures.
.1 " " " " " five " "
.01 " " " " " six " "

in all the work. If the final result obtained in this way conflicts with Rule 3, apply the latter.

Rule 6. *When logarithms are used, retain as many places in the mantissæ as there are significant figures retained in the data under Rule 5.*

The application of these rules is not always possible in the course of the work, since the precision measures may not be known until the end of the computation. But as a general rule it is sufficient in direct observations to retain one more place of figures than is given by the individual observations, and in indirect observations to retain two additional places.

CONDITIONED OBSERVATIONS.

33. Conditioned Observations are those in which the unknown quantities must be determined not only so as to satisfy as closely as possible the observation equations, but also so as to satisfy exactly certain other conditions. These conditions must be less in number than the unknown quantities, otherwise the unknowns could be determined from the conditions alone.

The adjustment of observations of this class may be reduced to the method already used for unconditioned observations in the following way.

The observations are represented by "Observation Equations," and the conditions by certain other equations, called "Condition Equations."

Between these two sets of equations we will eliminate as many unknowns as there are conditions. From the resulting equations, which will be the same in number as the observations, we will form in the usual manner the "Normal Equations" for the remaining unknowns. Having solved these normal equations and substituted the results in the condition equations, we shall obtain the values of the unknowns first eliminated.

All conditions of the problem are now fulfilled, for the condition equations are satisfied exactly and, moreover, according to the principle of Least Squares our results are those rendered most probable by the existence of the observations.

As in the example last considered, it is often more advantageous to compute corrections to the observed values of the unknown quantities rather than the values of the quantities themselves.

34. *Example.* Find the most probable values of the angles of a quadrilateral from the observations,

$$\begin{array}{rrrrl} A= & 101^\circ & 13' & 22'' & \text{weight } 3 \\ B= & 93 & 49 & 17 & \quad \text{"} \quad 2 \\ C= & 87 & 5 & 39 & \quad \text{"} \quad 2 \\ D= & 77 & 52 & 40 & \quad \text{"} \quad 1 \\ \hline A+B+C+D= & 360^\circ & 0' & 58'' & \end{array} \tag{A}$$

The condition to be satisfied is in this problem

$$A+B+C+D=360^\circ \tag{B}$$

Let z_1, z_2, z_3, z_4 be the most probable corrections to add to the observed values. This gives for

OBSERVATION EQUATIONS

$$\begin{array}{rl} z_1=0 & \text{weight } 3 \\ z_2=0 & \quad \text{"} \quad 2 \\ z_3=0 & \quad \text{"} \quad 2 \\ z_4=0 & \quad \text{"} \quad 1 \end{array} \tag{C}$$

and for the

CONDITION EQUATION

$$z_1+z_2+z_3+z_4+58=0 \tag{D}$$

Eliminating z_4 between (D) and (C), the equations from which the normal equations are to be derived become

$$\begin{array}{rl} z_1=0 & \text{weight } 3 \\ z_2=0 & \quad \text{"} \quad 2 \\ z_3=0 & \quad \text{"} \quad 2 \\ z_1+z_2+z_3+58=0 & \quad \text{"} \quad 1 \end{array} \tag{E}$$

Applying the rule in paragraph 27, these give the

NORMAL EQUATIONS

$$\begin{aligned} 4z_1 + z_2 + z_3 + 58 &= 0 \\ z_1 + 3z_2 + z_3 + 58 &= 0 \\ z_1 + z_2 + 3z_3 + 58 &= 0 \end{aligned} \qquad \text{(F)}$$

Solving, and substituting the results in equation (D), we find

$$\begin{aligned} z_1 &= -\ \ 8.29 & z_3 &= -12.43 \\ z_2 &= -12.43 & z_4 &= -24.85 \end{aligned} \qquad \text{(G)}$$

Applying these corrections to the observations (A), the most probable values of the angles are

$$\begin{aligned} A &= 101^\circ\ 13'\ 13''.71 \\ B &= \ \ 93\ \ \ 49\ \ \ \ 4.57 \\ C &= \ \ 87\ \ \ \ 5\ \ \ 26.57 \\ D &= \ \ 77\ \ \ 52\ \ \ 15.15 \end{aligned} \qquad \text{(H)}$$

Note. *In eliminating unknowns between the observation and condition equations care must be taken that the observation equations are not combined with each other or multiplied by any quantity. For if this is done the weights of the observation equations will be altered.* (See §28.)

35. In the above example it is evident that the corrections to be applied to the different observations are inversely as their weights. And, in general, when there is but one equation of condition, the observations expressing direct determinations of the unknowns, the corrections will be proportional to the coefficients of the unknowns in the equation of condition divided by the weights of the corresponding observations. A proof of this is given in paragraph **117**.

The most common case is that in which these coefficients are all unity, as in the example just solved, and we may then derive the

Rule. *Find the difference between the theoretical and observed results and divide this correction among the observations in the inverse ratio of their weights.*

In the last example the sum of the observed angles exceeds 360° by 58″. Therefore the correction to be applied to A is

$$-58 \times \frac{\frac{1}{3}}{\frac{1}{3}+\frac{1}{2}+\frac{1}{2}+1} = -58 \times \frac{1}{7} = -8.29$$

EMPIRICAL FORMULAS AND CONSTANTS.

36. In the work so far considered the observations are supposed to be made either directly upon the values of the unknown quantities or upon some function of the unknowns whose form, and the constants entering into it, are definitely known. But another sort of problem frequently occurs, in which observations are made upon the values of a certain variable and the corresponding values of some function of it, the exact form of the function not being known. The object in this case is the determination of the most probable form of the function and the values of the constants involved; that is, the derivation of the algebraic expression best representing the law connecting the variable and function.

This expression may be looked upon as the equation of a curve, abscissas denoting values of the variable and ordinates values of the function, and for all values of the variable within the range of the observations we may determine from it the most probable values of the function corresponding. But except in special cases, where the number of observations is large, where the law connecting variable and function is well defined, and where the equation obtained is an accurate representation of this law, it cannot be assumed to apply beyond the range of the observations. And in no case would it be safe to make use of the curve very far beyond the limits of the observations.

37. The Method of Least Squares will not assist in determining the form of the function. This must be settled upon beforehand, either from theoretical considerations or by constructing a plot, using values of the variable as abscissas and of the function as ordinates, when the smooth curve drawn through the points thus obtained will indicate the form of equation to be used.

It is to be observed that this is a method of trial and will not necessarily give the *most* probable form of the function; and in fact we may not be able to obtain the form that would be absolutely best. Further, several forms of equation may be known which would represent well the plotted points. In such a case that should be considered the best in which the sum of the squares of the residuals is found to be the least.

38. As soon as the form of the function is decided upon it should be reduced to the linear form, and the determination of the values of the constants involved is then a simple application of the preceding methods.

As the "Observation Equations" in any given problem will all be of the same kind, it is usually advisable to write out the general form of the "Normal Equations" and arrange the computations in tabular form, while the retention of the proper number of significant figures is of particular importance in this work.

39. A case that frequently occurs is that in which the quantity y is a constantly increasing function of the variable x, or where the plotted curve is approximately parabolic in form. Here the equation

$$y = A + Bx + Cx^2 + Dx^3 + \dots \qquad (27)$$

may be taken to represent the relation between the variable

and the function. The larger the number of terms taken in the second member, the more accurately may the equation obtained be made to represent the results of the observations; but the labor involved increases rapidly with increase in the number of terms, and if the plot shows a very nearly straight line the first two terms alone may suffice.

40. *Example.* In measuring the velocity of the current of a river the following results were obtained:—

Depths. x	Velocities. V
1	4.86
2	5.14
3	5.15
4	4.85
5	4.24
6	3.36
7	2.16
8	0.67

The velocity at the surface is 4.250. Find the equation of a curve which will express the relation between x and V.

Plotting the observations we find the curve

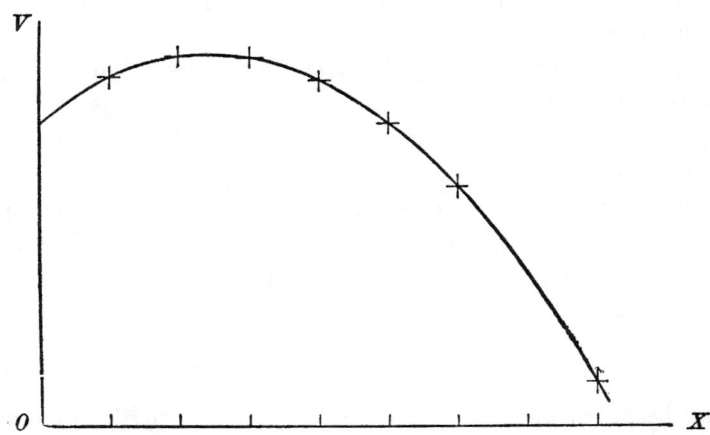

This is approximately parabolic in form and passes through the fixed point (0, 4.25). Therefore the relation between x and V may be expressed by the equation

$$V = 4.25 + Bx + Cx^2 \qquad \text{(A)}$$

and substituting in this the corresponding values of x and V as given by the observations, we shall have eight observation equations from which the most probable values of B and C are to be computed.

All of the observation equations being of the form (A) we have the

NORMAL EQUATIONS

$$C\,\Sigma x^4 + B\,\Sigma x^3 + 4.25\,\Sigma x^2 - \Sigma\,Vx^2 = 0 \qquad \text{(a)}$$

$$C\,\Sigma x^3 + B\,\Sigma x^2 + 4.25\,\Sigma x - \Sigma\,Vx = 0 \qquad \text{(b)}$$

For computing the coefficients in these equations it is most convenient to arrange the following table.

x	V	Vx	x^2	Vx^2	x^3	x^4
1	4.86	4.86	1	4.86	1	1
2	5.14	10.28	4	20.56	8	16
3	5.15	15.45	9	46.35	27	81
4	4.85	19.40	16	77.60	64	256
5	4.24	21.20	25	106.00	125	625
6	3.36	20.16	36	120.96	216	1296
7	2.16	15.12	49	105.84	343	2401
8	0.67	5.36	64	42.88	512	4096
36		111.83	204	525.05	1296	8772
Σx		$\Sigma\,Vx$	Σx^2	$\Sigma\,Vx^2$	Σx^3	Σx^4

Substituting these results in (a) and (b) we have

$$8772\,C + 1296\,B + 341.95 = 0 \qquad \text{(c)}$$

$$1296\,C + 204\,B + 41.17 = 0 \qquad \text{(d)}$$

and solving,

$$C = -.1493 \qquad B = .7465 \qquad \text{(e)}$$

Therefore the required equation is

$$V = 4.25 + .7465x - .1493x^2 \qquad \text{(f)}$$

Whenever the quantities in the observations are so large that the use of logarithms is desirable, these can best be put in the same columns directly over the natural numbers.

41. It is to be remarked that in solutions like the above we assume, from our method of forming the observation and normal equations, that the observations on the values of the function are alone subject to error, the observations on the variable being supposed to be exact or to have errors so small as to be negligible.

42. Periodic Phenomena. If as the variable increases the function passes through recurring values, that is, if y is a periodic function of x, some form of trigonometric equation would be the proper one to select. For instance, a good form to use is

$$y = A + B \sin \frac{360^\circ}{m} x + C \cos \frac{360^\circ}{m} x \qquad (28)$$

where A, B, C are the constants whose most probable values are to be found, and m is the number of units of x comprised in the entire cycle of values of y. The quantity m is to be determined from an inspection of the observations, and if it appears that several values of m might be used, all should be tried, and that which leads to the smallest sum for the squares of the residuals is to be considered the best. If the several cycles are not similar and regular, additional terms involving multiples of x will have to be added to equation (28).

43. Special Treatment of Exponential Equations. If the equation selected is not in the linear form as regards the unknown constants the general method of procedure is given in paragraph 44, but in some special cases a more simple reduction is possible. A quite common case is the following:—

Suppose the relation between x and y is expressed by the equation

$$y = kx^m \tag{29}$$

the problem being to obtain the best values of k and m. Taking the logarithms of both members of (29),

$$\log y = \log k + m \log x$$

Denoting $\log k$ by k', this equation becomes

$$m \log x + k' - \log y = 0$$

which is in the linear form as regards the unknowns m and k', and the normal equations may now be formed in the usual manner.

The most convenient way of determining whether equation (29) is a suitable one to select or not is to plot the corresponding values of x and y on *logarithmic* cross-section paper. If (29) is a proper equation to use the plotted points will lie upon a straight line.

REDUCTION OF OBSERVATION EQUATIONS TO THE LINEAR FORM.

44. When the observation equations are not linear as regards the unknowns, the only practicable method of procedure, as already mentioned in paragraph 23, is to reduce them to that form. This reduction may be effected as follows:—

Let the observation equations be

$$\left.\begin{aligned} f_1(Z_1, Z_2, \ldots Z_q) &= M_1 \\ f_2(Z_1, Z_2, \ldots Z_q) &= M_2 \\ \ldots\ldots & \\ \ldots\ldots & \\ f_n(Z_1\ Z_2, \ldots Z_q) &= M_n \end{aligned}\right\} \qquad \text{(A)}$$

in which $Z_1, Z_2, \ldots Z_q$ represent the unknown quantities and $M_1, M_2, \ldots M_n$ the observations, the functions being of known form.

Let $Z_1', Z_2', \ldots Z_q'$ be approximate values of $Z_1, Z_2, \ldots Z_q$, found by trial or the solution of a sufficient number of the observation equations, and let the most probable values of $Z_1, Z_2, \ldots Z_q$ be

$$Z_1' + z_1, \ Z_2' + z_2, \ldots Z_q' + z_q \qquad \text{(B)}$$

$z_1, z_2, \ldots z_q$ being small corrections whose values are to be determined by the Method of Least Squares. The first observation equation in (A) may then be written,

$$f_1(Z_1' + z_1, Z_2' + z_2, \ldots Z_q' + z_q) = M_1$$

Expanding the first member by Taylor's Theorem, denoting $f_1(Z_1', Z_2', \ldots Z_q')$ by k_1, and neglecting terms containing powers of $z_1, z_2, \ldots z_q$ higher than the first, this becomes

$$k_1 + \frac{\partial k_1}{\partial Z_1'} z_1 + \frac{\partial k_1}{\partial Z_2'} z_2 + \ldots \frac{\partial k_1}{\partial Z_q'} z_q = M_1$$

If now we represent the coefficients of $z_1, z_2, \ldots z_q$, by $a_1, b_1, \ldots q_1$, also $k_1 - M_1$ by m_1, and treat the other equations in (A) in the same manner, our observation equations will take the form

$$
\begin{aligned}
a_1 z_1 + b_1 z_2 + \ldots q_1 z_q + m_1 &= v_1 \\
a_2 z_1 + b_2 z_2 + \ldots q_2 z_q + m_2 &= v_2 \\
\ldots \quad \ldots \quad \ldots & \\
\ldots \quad \ldots \quad \ldots & \\
a_n z_1 + b_n z_2 + \ldots q_n z_q + m_n &= v_n
\end{aligned} \tag{C}
$$

The second members reduce to the residuals since the quantities $z_1, z_2, \ldots z_q$ represent merely the most probable values of the corrections. Equations (C) can now be solved in the usual manner and the most probable values of $Z_1, Z_2, \ldots Z_q$ found by substituting the results in (B).

45. *Example.* From the following observations find the most probable values of x and y.

$$
\begin{aligned}
\sin x + \cos 2y &= 1.5 \\
\cos x + 3 \sin y &= 1.7 \\
x^2 + 5y &= 2.1
\end{aligned} \tag{A}
$$

By trial it is found that 42° and 18° are approximate values of x and y. Then in accordance with paragraph 44 we put

$$x = 42° + z_1 \qquad y = 18° + z_2 \tag{B}$$

and expanding the different functions, we find

$$
\begin{aligned}
k_1 &= \sin 42° + \cos 36°, & \frac{\partial k_1}{\partial x} &= \cos 42°, & \frac{\partial k_1}{\partial y} &= -2 \sin 36° \\
&= 1.48 & &= .74 & &= -1.18 \\
k_2 &= \cos 42° + 3 \sin 18°, & \frac{\partial k_2}{\partial x} &= -\sin 42°, & \frac{\partial k_2}{\partial y} &= 3 \cos 18° \\
&= 1.67 & &= -.67 & &= 2.85 \\
k_3 &= \left(\frac{42\pi}{180}\right)^2 + 5\,\frac{18\pi}{180}, & \frac{\partial k_3}{\partial x} &= 2\,\frac{42\pi}{180}, & \frac{\partial k_3}{\partial y} &= 5 \\
&= 2.11 & &= 1.47 & &= 5
\end{aligned}
$$

$$
\begin{aligned}
k_1 - M_1 &= -.02, & k_2 - M_2 &= -.03, & k_3 - M_3 &= .01 \\
&= m_1, & &= m_2, & &= m_3
\end{aligned}
$$

Therefore we have for

OBSERVATION EQUATIONS

$$\begin{aligned} .74z_1 - 1.18z_2 - .02 &= v_1 \\ -.67z_1 + 2.85z_2 - .03 &= v_2 \\ 1.47z_1 + 5z_2 + .01 &= v_3 \end{aligned} \qquad \text{(C)}$$

From these are obtained the

NORMAL EQUATIONS

$$\begin{aligned} 3.16z_1 + 4.57z_2 + .020 &= 0 \\ 4.57z_1 + 34.51z_2 - .012 &= 0 \end{aligned} \qquad \text{(D)}$$

Solving,

$$z_1 = -.00845 \qquad z_2 = .00147 \qquad \text{(E)}$$

These results are in circular measure. Reducing to degree measure we have

$$z_1 = -29'.0 \qquad z_2 = 5'.1 \qquad \text{(F)}$$

Substituting these in (B),

$$x = 41^\circ\ 31'.0 \qquad y = 18^\circ\ 5'.1 \qquad \text{(G)}$$

46. If the observations are conditioned, precisely the same method will be followed in reducing all the equations to the linear form. The rest of the solution will then be as usual.

If the values found for $z_1, z_2, \ldots z_q$ should turn out to be so large that the terms involving their second and higher powers can not be neglected as assumed, the process must be repeated using the values of $Z_1, Z_2, \ldots Z_q$ first obtained as approximations.

In a few cases the equations may be reduced to the linear form by some special artifice of a simple character, as in paragraph 43. In such cases the method of expansion by Taylor's Theorem should not be used.

CHAPTER III.

THE PRECISION OF OBSERVATIONS.

47. The work so far considered has treated solely of the methods by means of which the most probable values of the unknown quantities may be determined from a series of observations. But in general something more than this is desired. We wish to know, if possible, how much reliance can be placed upon the results obtained, and how they compare in precision with other determinations of the same quantities. Preliminary to the discussion of the precision of our results it will be necessary to consider more fully the Curve of Error.

48. The Constant *k*. The law of distribution of errors of observation has been shown to be represented by a curve whose equation is

$$y = ke^{-h^2x^2}$$

In this, if $x = 0, \quad y = k \qquad (30)$

Therefore the constant k represents the intercept of the Curve of Error on the axis of Y. It is not, however, an independent quantity but is determined by the value of h, as will now be shown.

49. To Find *k* in Terms of *h*. Since, as shown in paragraph 18, the total area between the curve and the axis of X is denoted by unity, we have

$$k\int_{-\infty}^{\infty} e^{-h^2x^2}\,dx = 1 \quad \text{or} \quad k\int_{0}^{\infty} e^{-h^2x^2}\,dx = \frac{1}{2} \qquad \text{(a)}$$

This may be written

$$\int_{0}^{\infty} e^{-h^2x^2}\,h dx = \frac{h}{2k} \qquad \text{(b)}$$

Let $t = hx$, $\therefore\ dt = hdx$. Also, when $x = \infty$, $t = \infty$, and when $x = 0$, $t = 0$.

$$\therefore \qquad \int_0^\infty e^{-t^2}\,dt = \int_0^\infty e^{-h^2x^2}\,h\,dx \tag{c}$$

Multiplying this equation by

$$\int_0^\infty e^{-t^2}\,dt = \int_0^\infty e^{-h^2}\,dh$$

we have

$$\left[\int_0^\infty e^{-t^2}\,dt\right]^2 = \int_0^\infty\int_0^\infty e^{-h^2(1+x^2)}\,h\,dx\,dh$$

$$= \int_0^\infty -\frac{1}{2(1+x^2)}\,dx \int_0^\infty e^{-h^2(1+x^2)}(-2h)(1+x^2)\,dh$$

$$= \frac{1}{2}\int_0^\infty \frac{dx}{1+x^2}$$

$$= \frac{1}{2}\left[\tan^{-1} x\right]_0^\infty = \frac{\pi}{4}$$

$$\therefore \qquad \int_0^\infty e^{-t^2}dt = \frac{\sqrt{\pi}}{2} \tag{31}$$

Substituting (31) in (c) and (b)

$$\frac{\sqrt{\pi}}{2} = \frac{h}{2k} \qquad \text{or} \qquad k = \frac{h}{\sqrt{\pi}} \tag{32}$$

Therefore equations (16) and (17) become

$$\left.\begin{aligned} y &= \frac{h}{\sqrt{\pi}}\,e^{-h^2v^2} \\ y &= \frac{h}{\sqrt{\pi}}\,e^{-h^2x^2} \end{aligned}\right\} \tag{33}$$

50. It has been shown that the quantity h is a "Measure of Precision" of the observations, and hence a determination of its value in each case would enable us to compare the relative reliability of different measurements. In practice, however, it is not found convenient to compute the value of h directly, and so this quantity is used only in developing the theory of the subject, while in the comparison of observations certain other quantities now to be derived are used as precision measures. These latter quantities are called the "Mean of the Errors," the "Mean Error," and the "Probable Error," respectively, and may be computed directly from the observations. Further, as it will be found that they all bear a definite relation to h, the value of this quantity can be determined from them if desired.

In the following discussions no distinction is made between positive and negative errors of the same numerical magnitude, and unless otherwise stated the observations are all of the same weight.

THE MEAN OF THE ERRORS OR AVERAGE DEVIATION.

51. The Average Deviation (a.d.) of an observation is the arithmetical mean of the errors all taken with the positive sign.

Since from (6) the probability that the error of a single observation will fall between x and $x + dx$ is

$$\phi(x)\,dx,$$

if n observations are made, the number of errors falling between these limits is

$$n\,\phi(x)\,dx.$$

Hence the sum of all the errors of the observations is

$$n\int_{-\infty}^{\infty} x\,\phi(x)\,dx \quad \text{or} \quad 2n\int_{0}^{\infty} x\,\phi(x)\,dx.$$

Dividing this last expression through by n we have

$$a.d. = 2\int_0^\infty x\,\phi(x)\,dx = \frac{2h}{\sqrt{\pi}}\int_0^\infty e^{-h^2x^2}\,x\,dx$$

$$= -\frac{1}{h\sqrt{\pi}}\int_0^\infty e^{-h^2x^2}(-2h^2x)\,dx$$

or

$$\boldsymbol{a.d.} = \frac{1}{h\sqrt{\pi}} \qquad \textbf{(34)}$$

THE MEAN ERROR.

52. The Mean Error (μ) of an observation is the square root of the arithmetical mean of the squares of the errors.

The total number of errors being n, the number falling between x and $x + dx$ is, as just shown,

$$n\,\phi(x)\,dx,$$

and the sum of the squares of these errors is

$$n\,x^2\,\phi(x)\,dx.$$

Therefore the sum of the squares of all the errors is

$$n\int_{-\infty}^{\infty} x^2\,\phi(x)\,dx$$

$$\therefore \quad \mu^2 = \frac{h}{\sqrt{\pi}}\int_{-\infty}^{\infty} e^{-h^2x^2}\,x^2\,dx \qquad \text{(a)}$$

But as shown in paragraph 49,

$$\frac{h}{\sqrt{\pi}}\int_{-\infty}^{\infty} e^{-h^2x^2}\,dx = 1 \quad \text{or} \quad \int_{-\infty}^{\infty} e^{-h^2x^2}\,dx = \frac{\sqrt{\pi}}{h} \qquad \text{(b)}$$

Differentiating (b) with respect to h,

$$-2h\int_{-\infty}^{\infty} e^{-h^2x^2}\,x^2\,dx = -\frac{\sqrt{\pi}}{h^2} \qquad \text{(c)}$$

and replacing the integral in (a) by its value as determined from equation (c), we have

$$\mu = \frac{1}{h\sqrt{2}} \tag{35}$$

THE PROBABLE ERROR.

53. The Probable Error (r) of an observation is an error such that one-half the errors of the series are greater than it and the other half less than it. Or it is an error of such a magnitude that the probability of making an error greater than it in any given observation is just equal to the probability of making one less than it, both probabilities being one-half.

The probability that the error of an observation will fall between x and $x + dx$ being $\phi(x)\,dx$, the probability that the error will fall between the limits r and $-r$ is

$$P = \int_{-r}^{r} \phi(x)\,dx = \frac{2h}{\sqrt{\pi}} \int_0^r e^{-h^2x^2}\,dx \tag{36}$$

If r is the probable error, P is one-half, or

$$\frac{2h}{\sqrt{\pi}} \int_0^r e^{-h^2x^2}\,dx = \frac{1}{2} \tag{37}$$

and from this definite integral r is to be found.

Let $t = hx$, $\therefore$ $dt = hdx$. Also when $x = r$, we have $t = hr$, and when $x = 0$, $t = 0$. Substituting these results in (37), we have

$$\frac{2}{\sqrt{\pi}} \int_0^{hr} e^{-t^2}\,dt = \frac{1}{2} \tag{38}$$

63. *Example.* Given a series of observations on M, the difference in longitude between two stations. To find M_0, μ, μ_0, *a.d.*, *A.D.*, r, r_4, r_0.

M	p	pM	v	$v\sqrt{p}$	v^2	pv^2
1° 4′ 30″	4	120	1.6	3.2	2.6	10.4
41	1	41	9.4	9.4	88.4	88.4
43	1	43	11.4	11.4	130.0	130.0
37	9	333	5.4	16.2	29.1	261.9
48	4	192	16.4	32.8	269.0	1076.0
34	16	544	2.4	9.6	5.7	91.2
25	9	225	6.6	19.8	43.6	392.4
46	1	46	14.4	14.4	207.4	207.4
28	25	700	3.6	18.0	13.0	325.0
24	4	96	7.6	15.2	57.7	230.8
1° 4′ 31″.6	74	2340		150.0		2813.5
M_0	Σp	Σpm		$\Sigma v\sqrt{p}$		Σpv^2

By (3)
$$M_0 = \frac{\Sigma pM}{\Sigma p} = 1^\circ\, 4' + \frac{2340''}{74}$$
$$= 1^\circ\, 4'\, 31''.6$$

$$\mu = \sqrt{\frac{2813.5}{9}} = 17.7 \qquad \mu_0 = \frac{\mu}{\sqrt{74}} = 2.1$$

$$r = .6745\,\mu = 11.9 \qquad r_4 = \frac{r}{3} = 4.0$$

$$a.d. = \frac{150}{\sqrt{90}} = 15.8 \qquad A.D. = \frac{a.d.}{\sqrt{74}} = 1.8$$

$$r_0 = .6745\,\mu_0 = 1.4 = .8453\,A.D. = 1.5$$

$$\therefore \quad M_0 = 1^\circ\, 4'\, 31''.\,6 \pm 1''.4$$

FUNCTIONS OF OBSERVED QUANTITIES.

64. Theorem. Given any number of quantities and their Mean and Probable Errors and Average Deviations, to find the Mean and Probable Errors and Average Deviation of any function of the quantities.

Let the quantities be $M_1, M_2, \ldots M_q$.
" " mean errors be $\mu_1, \mu_2, \ldots \mu_q$.
" " function be $M = f(M_1, M_2, \ldots M_q)$.
" " mean error of M be E.
" " probable error of M be R.
" " average deviation of M be D.

The derivation of the general formula will be simplified if we consider first a few special forms of functions.

65. *Case I.* Suppose $M = M_1 \pm M_2$.

The number of observations from which M_1 and M_2, and hence μ_1 and μ_2, have been determined is not necessarily known, but we may assume that for each quantity it is any large number n, and that the real errors of the observations are

for M_1, $x_1', x_1'', x_1''', \ldots$
" M_2, $x_2', x_2'', x_2''', \ldots$

Then the real errors of M, computed from the separate observations on M_1 and M_2 will be

$$x_1' \pm x_2', \quad x_1'' \pm x_2'', \quad x_1''' \pm x_2''' \ldots$$

$$\therefore \quad E^2 = \frac{(x_1' \pm x_2')^2 + (x_1'' \pm x_2'')^2 \ldots}{n}$$

$$= \frac{\Sigma x_1^2 \pm 2\,\Sigma x_1 x_2 + \Sigma x_2^2}{n}$$

or

$$E^2 = \mu_1^2 + \mu_2^2 \qquad (64)$$

since in the most probable case the term $\Sigma x_1 x_2$ will disappear, as there most likely will be as many positive as negative products of the same absolute magnitude of the form $x_1 x_2$.

By successive applications of the above the same principle may be extended to cover the algebraic sum of any number of quantities. So that if

$$M = M_1 \pm M_2 \pm \ldots M_q$$

then $$E^2 = \mu_1^2 + \mu_2^2 + \ldots \mu_q^2 = \Sigma \mu^2 \qquad (65)$$

Since probable errors and average deviations differ from mean errors merely by a constant factor, we shall likewise have

$$R^2 = r_1^2 + r_2^2 + \ldots r_q^2 = \Sigma r^2 \qquad (66)$$

$$D^2 = a.d._1^2 + a.d._2^2 + \ldots = \Sigma a.d.^2 \qquad (67)$$

66. *Example.* Given the telegraphic longitude results,

		h.	m.	sec.	sec.
(A)	Cambridge west of Greenwich,	4	44	30.99	± 0.23
(B)	Omaha " " Cambridge,	1	39	15.04	± 0.06
(C)	Springfield east " Omaha,		25	8.69	± 0.11

Find L, the longitude of Springfield, and its probable error.

$$L = A + B - C = 5^h\ 58^m\ 37^s.34 \pm 0^s.26$$

for by (66) $$R = \sqrt{(.23^2 + .06^2 + .11^2)} = .26$$

67. *Case II.* Suppose $M = a_1 M_1$

Using the same notation as in paragraph 64, the real errors of M will be

$$a_1 x_1',\ a_1 x_1'',\ a_1 x_1''',\ \ldots$$

$$\therefore \qquad E^2 = \frac{a_1^2\, \Sigma x_1^2}{n} = a_1^2 \mu_1^2$$

or $$E = a_1 \mu_1 \qquad (68)$$

68. *Case III.* Suppose $M = a_1 M_1 \pm a_2 M_2 \ldots a_q M_q$

By combining (65) and (68)

$$E^2 = \Sigma a^2 \mu^2$$

also $$R^2 = \Sigma a^2 r^2 \qquad (69)$$

and $$D^2 = \Sigma a^2 a.d.^2$$

69. *Example.* The length of a bar at 20° Centigrade is found to be 75″.0041 ± 0″.0037, and its expansion per centigrade degree 0″.0036 ± 0″.0018. What is the length of the bar at 56° Fahrenheit?

$$20° \text{ C.} = \frac{9}{5} \times 20 + 32 = 68° \text{ F.}$$

The expansion for 1° F. will be $\frac{5}{9} \times .0036 \pm \frac{5}{9} \times .0018$

or $$.0020 \pm .0010$$

$$\therefore \quad L = 75.0041 - 12 \times .0020$$

$$= 74.9801$$

$$R = \sqrt{.0037^2 + (12 \times .001)^2}$$

$$= .013$$

$$\therefore \quad L = 74''.980 \pm 0''.013$$

70. *Case IV.* Suppose $M = f(M_1, M_2, \ldots M_q)$.

Let $M_1 = a_1 + m_1$, $M_2 = a_2 + m_2$, ... $M_q = a_q + m_q$, where $a_1, a_2, \ldots a_q$ are arbitrarily assumed quantities very nearly equal to $M_1, M_2, \ldots M_q$, so that $m_1, m_2, \ldots m_q$ are so small that their second and higher powers may be neglected. Then the errors of $M_1, M_2, \ldots M_q$ may be considered to be in $m_1, m_2, \ldots m_q$, and hence $\mu_1, \mu_2, \ldots \mu_q$

may be regarded as the mean errors of the quantities $m_1, m_2, \ldots m_q$. We now have

$$M = f(a_1 + m_1, a_2 + m_2, \ldots a_q + m_q)$$

Expanding this expression by Taylor's Theorem, and denoting $f(a_1, a_2, \ldots a_q)$ by M', we have

$$M = M' + m_1 \frac{\partial M'}{\partial a_1} + m_2 \frac{\partial M'}{\partial a_2} + \ldots m_q \frac{\partial M'}{\partial a_q} \quad \text{(A)}$$

+ negligible terms in the second and higher powers of $m_1, m_2, \ldots m_q$.

Then the mean error of M will be the same as the mean error of terms (A), and by (69) this is given by

$$E^2 = \mu_1^2 \left[\frac{\partial M'}{\partial a_1}\right]^2 + \ldots . \mu_q^2 \left[\frac{\partial M'}{\partial a_q}\right]^2$$

or what is practically the same thing, by

$$\boldsymbol{E}^2 = \boldsymbol{\mu}_1^2 \left[\frac{\partial \boldsymbol{M}}{\partial \boldsymbol{M}_1}\right]^2 + \ldots . \boldsymbol{\mu}_q^2 \left[\frac{\partial \boldsymbol{M}}{\partial \boldsymbol{M}_q}\right]^2$$

$$\therefore \quad \boldsymbol{R}^2 = \boldsymbol{r}_1^2 \left[\frac{\partial \boldsymbol{M}}{\partial \boldsymbol{M}_1}\right]^2 + \ldots . \boldsymbol{r}_q^2 \left[\frac{\partial \boldsymbol{M}}{\partial \boldsymbol{M}_q}\right]^2 \quad \textbf{(70)}$$

$$\boldsymbol{D}^2 = \boldsymbol{a.d.}_1^2 \left[\frac{\partial \boldsymbol{M}}{\partial \boldsymbol{M}_1}\right]^2 + \ldots \boldsymbol{a.d.}_q^2 \left[\frac{\partial \boldsymbol{M}}{\partial \boldsymbol{M}_q}\right]^2$$

71. *Example A.* Two sides of a right triangle are measured with results

$$a = 49.53 \pm 0.59$$
$$b = 50.38 \pm 0.93$$

Find the length of the hypotenuse c and its probable error. In this example

$$M = \sqrt{M_1^2 + M_2^2} = \sqrt{a^2 + b^2} = 70.65$$

$$\therefore \quad \frac{\partial M}{\partial M_1} = \frac{a}{\sqrt{a^2 + b^2}}, \quad \frac{\partial M}{\partial M_2} = \frac{b}{\sqrt{a^2 + b^2}}$$

By (70) $\quad R^2 = r_1^2 \dfrac{a^2}{a^2 + b^2} + r_2^2 \dfrac{b^2}{a^2 + b^2}$

$$= .59^2 \left[\frac{49.53}{70.65}\right]^2 + .93^2 \left[\frac{50.38}{70.65}\right]^2$$

$$\therefore \quad R = .78$$

and $$c = 70.65 \pm 0.78$$

Example B. If the probable error of x is r, what is the probable error of the common logarithm of x? In this case

$$M = \log_{10} M_1 = \log_{10} x$$

$$\frac{\partial M}{\partial M_1} = \frac{\partial \log_{10} x}{\partial x} = \frac{\log_{10} e}{x}$$

$$\therefore \quad R = \frac{\log_{10} e}{x} r$$

Example C. If the weight of x is p, what is the weight p_0 of sin x?

Denoting the mean errors of x and sin x by μ and E, respectively, we have, by (45) and (70),

$$E^2 = \mu^2 \left[\frac{\partial \sin x}{\partial x}\right]^2 = \mu^2 \cos^2 x$$

and
$$\frac{p_0}{p} = \frac{\mu^2}{E^2}$$

$$\therefore \qquad p_0 = \frac{p}{\cos^2 x} = p \sec^2 x$$

72. Equation (70), which expresses the law of propagation of error in functions of observed quantities, is one of the most important in the whole theory of the Method of Least Squares. Upon it in particular is based the discussion of the "Precision of Measurements." This subject treats, in the first place, of the methods of finding the precision of a quantity obtained by computation from a series of measured quantities; and, in the second place, it investigates the precision with which the component measurements of a series must be made in order to obtain a required degree of precision in the final result. The following simple example will illustrate the character of the solutions:—

73. *Example.* In the determination of a current by a tangent galvanometer we have

$$I = 10 \frac{H}{G} \tan \phi$$

where I is the current in amperes, H the horizontal component of the earth's magnetic force, G the galvanometer constant, and ϕ the angle of deflection. Given the errors δ_1, δ_2, δ_3, in H, G and $\tan \phi$, to find the error Δ in I.

By (70)

$$\Delta^2 = \delta_1^2 \left[\frac{\partial I}{\partial H}\right]^2 + \delta_2^2 \left[\frac{\partial I}{\partial G}\right]^2 + \delta_3^2 \left[\frac{\partial I}{\partial \tan \phi}\right]^2$$

$$= \frac{100}{G^2} \delta_1^2 \tan^2 \phi + \frac{100 H^2}{G^4} \delta_2^2 \tan^2 \phi + \frac{100 H^2}{G^2} \delta_3^2 \qquad \text{(a)}$$

Dividing this equation by $I^2 = \frac{100\,H^2}{G^2}\tan^2\phi$ we have

$$\left[\frac{\Delta}{I}\right]^2 = \left[\frac{\delta_1}{H}\right]^2 + \left[\frac{\delta_2}{G}\right]^2 + \left[\frac{\delta_3}{\tan\phi}\right]^2 \qquad \text{(b)}$$

That is, the square of the percentage error in I is equal to the sum of the squares of the percentage errors in H, G, and $\tan\phi$. Hence if

H is determined within .4 per cent.
G " " " .2 " "
$\tan\phi$ " " " .1 " "

then
$$\frac{\Delta}{I} = \sqrt{.16 + .04 + .01} = .46 \text{ per cent.} \qquad \text{(c)}$$

Next, suppose the value of I is required to within .1 per cent. To find the necessary accuracy in the determinations of H, G, and $\tan\phi$ when the error in each of these quantities is to have the same influence upon the total error.

From (b) we shall now have

$$.001^2 = 3\left[\frac{\delta_1}{H}\right]^2 = 3\left[\frac{\delta_2}{G}\right]^2 = 3\left[\frac{\delta_3}{\tan\phi}\right]^2 \qquad \text{(d)}$$

$$\therefore \qquad \frac{\delta_1}{H} = \frac{.001}{\sqrt{3}} = .000577$$

$$\therefore \qquad \begin{aligned} \delta_1 &= .00058\,H \\ \delta_2 &= .00058\,G \\ \delta_3 &= .00058\tan\phi \end{aligned} \qquad \text{(e)}$$

It is comparatively easy to obtain the necessary accuracy in the measurements of G and $\tan\phi$, but difficult in the case of H.

For additional work of this kind see Holman's "Precision of Measurements."

74. Combination of Functions of the Same Variables. It is to be noticed that equation (70) applies only when M is a function of independent quantities. If $M_1, M_2, \ldots M_q$ are merely different functions of the same quantities we must proceed as follows:—

Let
$$M_1 = \phi(z_1, z_2, \ldots z_k)$$
$$M_2 = \psi(z_1, z_2, \ldots z_k)$$
$$M = f(M_1, M_2)$$

If any single observations of $z_1, z_2, \ldots z_k$ are subject to errors $x_1, x_2, \ldots x_k$, the corresponding errors in M_1 and M_2 will be

for M_1, $$X_1 = a_1x_1 + a_2x_2 + \ldots a_kx_k \qquad \text{(a)}$$
" M_2, $$X_2 = a_1'x_1 + a_2'x_2 + \ldots a_k'x_k \qquad \text{(b)}$$

Where $a_1, a_2, \ldots a_k$ are the differential coefficients of M_1, and $a_1', a_2' \ldots a_k'$ the differential coefficients of M_2 with respect to $z_1, z_2, \ldots z_k$. The corresponding error in M will then be

$$X = AX_1 + A'X_2 \qquad \text{(c)}$$

Where A and A' are the differential coefficients of M with respect to M_1 and M_2. Substituting in (c) from (a) and (b),

$$X = (Aa_1 + A'a'_1)\,x_1 + (Aa_2 + A'a'_2)\,x_2 \ldots$$
$$= \alpha x_1 + \beta x_2 + \ldots \lambda x_k$$

Then if the number of observations or values of X be denoted by n,

$$E^2 = \frac{\Sigma X^2}{n} = \alpha^2 \frac{\Sigma x_1^2}{n} + \beta^2 \frac{\Sigma x_2^2}{n} + \ldots$$
$$+ 2\alpha\beta \frac{\Sigma x_1x_2}{n} + \ldots$$

$$\therefore \quad E^2 = \alpha^2\mu_1^2 + \beta^2\mu_2^2 + \ldots \lambda^2\mu_k^2 \qquad \text{(d)}$$

since in the most probable case the product terms will cancel out.

Expanding (d) we have

$$E^2 = (Aa_1 + A'a_1')^2 \mu_1^2 + (Aa_2 + A'a_2')^2 \mu_2^2 + \dots$$
$$= A^2(a_1^2\mu_1^2 + a_2^2\mu_2^2 \dots) + A'^2(a_1'^2\mu_1^2 + a_2'^2\mu_2^2 \dots)$$
$$+ 2AA'(a_1a_1'\mu_1^2 + a_2a_2'\mu_2^2 + \dots) \qquad \textbf{(71)}$$

75. *Example.* As a very simple problem take

$$M_1 = 2z_1, \qquad M_2 = 3z_1, \qquad \mu_1 = 0.1,$$

and $$M = M_1 + M_2.$$

Then $A = 1, \quad A' = 1, \quad a_1 = 2, \quad a_1' = 3.$

By (71) $E^2 = 4 \times .01 + 9 \times .01 + 2 \times 2 \times 3 \times .01 = .25$

or $$E = 0.5$$

In this particular example the result may be found directly from (68) by substituting at first in M the values of M_1 and M_2.

Thus $$M = 2z_1 + 3z_1 = 5z_1$$
$$\therefore \quad E = 5\mu_1 = 0.5$$

If M_1 and M_2 had been independent quantities, by (64) or (69) we should have had

$$E = \sqrt{(2 \times 0.1)^2 + (3 \times 0.1)^2}$$
$$= 0.36$$

INDIRECT OBSERVATIONS.

76. The determination of the precision measures of the unknown quantities in case the observations are indirect involves a knowledge of the weights of the unknowns, and consequently the method of computing these weights must first be demonstrated. It will be assumed at present that all the observations are of weight unity.

FIRST METHOD OF COMPUTING THE WEIGHTS.

Let the observations be

$$\begin{array}{l} a_1 z_1 + b_1 z_2 + \ldots q_1 z_q + k_1 = M_1 \\ a_2 z_1 + b_2 z_2 + \ldots q_2 z_q + k_2 = M_2 \\ \quad\ldots\quad\ldots\quad\ldots \\ \quad\ldots\quad\ldots\quad\ldots \\ a_n z_1 + b_n z_2 + \ldots q_n z_q + k_n = M_n \end{array}$$

in which $M_1, M_2, \ldots M_n$ denote the actual observations, and $z_1, z_2, \ldots z_q$ the most probable values of the unknown quantities. Let

$$k_1 - M_1 = m_1, \quad k_2 - M_2 = m_2, \ldots k_n - M_n = m_n$$

Then the above equations give rise to the

OBSERVATION EQUATIONS

$$\begin{array}{l} a_1 z_1 + b_1 z_2 + \ldots q_1 z_q + m_1 = v_1 \\ a_2 z_1 + b_2 z_2 + \ldots q_2 z_q + m_2 = v_2 \\ \quad\ldots\quad\ldots\quad\ldots \\ \quad\ldots\quad\ldots\quad\ldots \\ a_n z_1 + b_n z_2 + \ldots q_n z_q + m_n = v_n \end{array} \qquad \text{(A)}$$

By the rule, paragraph 25, we now form the

NORMAL EQUATIONS

$$\begin{array}{l} z_1 \Sigma a^2 + z_2 \Sigma ab + \ldots z_q \Sigma aq + \Sigma am = 0 \\ z_1 \Sigma ab + z_2 \Sigma b^2 + \ldots z_q \Sigma bq + \Sigma bm = 0 \\ \quad\ldots\quad\ldots\quad\ldots \\ \quad\ldots\quad\ldots\quad\ldots \\ z_1 \Sigma aq + z_2 \Sigma bq + \ldots z_q \Sigma q^2 + \Sigma qm = 0 \end{array} \qquad \text{(B)}$$

Multiply the first of (B) by Q_1, the second of (B) by Q_2, ... and add the results. Then

$$
\begin{aligned}
& z_1\,(Q_1\Sigma\, a^2 + Q_2\Sigma\, ab + \ldots\, Q_q\Sigma\, aq) \\
& + z_2\,(Q_1\Sigma\, ab + Q_2\Sigma\, b^2 + \ldots\, Q_q\Sigma\, bq) \\
& \quad \ldots \qquad \ldots \qquad \ldots \\
& \quad \ldots \qquad \ldots \qquad \ldots \\
& + z_q(Q_1\Sigma\, aq + Q_2\Sigma\, bq + \ldots\, Q_q\Sigma\, q^2) \\
& + Q_1\Sigma\, am + Q_2\Sigma\, bm + \ldots\, Q_q\Sigma\, qm \\
& = 0
\end{aligned}
\qquad (C)
$$

Let $Q_1, Q_2, \ldots Q_q$ be determined so that the coefficient of z_1 in (C) shall be unity, and the coefficients of z_2, $z_3, \ldots z_q$ each equal to zero. That is, let

$$
\begin{aligned}
& Q_1\Sigma\, a^2 + Q_2\Sigma\, ab + \ldots\, Q_q\Sigma\, aq - 1 = 0 \\
& Q_1\Sigma\, ab + Q_2\Sigma\, b^2 + \ldots\, Q_q\Sigma\, bq = 0 \\
& \quad \ldots \qquad \ldots \qquad \ldots \\
& \quad \ldots \qquad \ldots \qquad \ldots \\
& Q_1\Sigma\, aq + Q_2\Sigma\, bq + \ldots\, Q_q\Sigma\, q^2 = 0
\end{aligned}
\qquad (D)
$$

Then (C) becomes

$$
z_1 + Q_1\Sigma\, am + Q_2\Sigma\, bm + \ldots\, Q_q\Sigma\, qm = 0 \qquad (E)
$$

Equations (D) may be derived from the normal equations (B), if in them we replace $z_1, z_2, \ldots z_q$ by $Q_1, Q_2, \ldots Q_q$, $\Sigma\, am$ by -1, and $\Sigma\, bm$, $\Sigma\, cm$, ... $\Sigma\, qm$ by zero. Hence the solution of the normal equations with these changes will give the values of $Q_1, Q_2, \ldots Q_q$.

TO SHOW THAT Q_1 IS THE RECIPROCAL OF THE WEIGHT OF z_1.

Expanding the coefficients of (E) we have

$$
\begin{aligned}
z_1 &+ Q_1(a_1 m_1 + a_2 m_2 + \ldots a_n m_n) \\
&+ Q_2(b_1 m_1 + b_2 m_2 + \ldots b_n m_n) \\
&\quad \ldots \qquad \ldots \qquad \ldots \\
&\quad \ldots \qquad \ldots \qquad \ldots \\
&+ Q_q(q_1 m_1 + q_2 m_2 + \ldots q_n m_n) \\
&= 0
\end{aligned}
$$

or collecting the coefficients of $m_1, m_2, \ldots m_n$

$$
\begin{aligned}
z_1 &+ m_1(Q_1 a_1 + Q_2 b_1 + \ldots Q_q q_1) \\
&+ m_2(Q_1 a_2 + Q_2 b_2 + \ldots Q_q q_2) \\
&\quad \ldots \qquad \ldots \qquad \ldots \\
&\quad \ldots \qquad \ldots \qquad \ldots \\
&+ m_n(Q_1 a_n + Q_2 b_n + \ldots Q_q q_n) \\
&= 0
\end{aligned}
\tag{E'}
$$

For convenience in writing we will let

$$
\begin{aligned}
\alpha_1 &= Q_1 a_1 + Q_2 b_1 + \ldots Q_q q_1 \\
\alpha_2 &= Q_1 a_2 + Q_2 b_2 + \ldots Q_q q_2 \\
&\quad \ldots \qquad \ldots \qquad \ldots \\
&\quad \ldots \qquad \ldots \qquad \ldots \\
\alpha_n &= Q_1 a_n + Q_2 b_n + \ldots Q_q q_n
\end{aligned}
\tag{F}
$$

$$
\therefore \qquad z_1 + \alpha_1 m_1 + \alpha_2 m_2 + \ldots \alpha_n m_n = 0 \tag{G}
$$

Multiply the first of (F) by a_1, the second by $a_2, \ldots$ and add the results. Then

$$
\begin{aligned}
\Sigma\, a\alpha &= Q_1 \Sigma\, a^2 + Q_2 \Sigma\, ab + \ldots Q_q \Sigma\, aq \\
&= 1 \qquad \text{by first of (D)}
\end{aligned}
\tag{H}
$$

Multiply the first of (F) by b_1, the second by $b_2, \ldots$ and add the results. Then

$$\Sigma\, ba = Q_1\Sigma\, ab + Q_2\Sigma\, b^2 + \ldots Q_q\Sigma\, bq$$
$$= 0 \qquad \text{by second of (D)}$$

Likewise $\qquad \Sigma\, ca = 0 \qquad$ (H′)

$$\ldots$$
$$\ldots$$
$$\Sigma\, qa = 0$$

Multiply the first of (F) by a_1, the second by $a_2, \ldots$ and add the results. This gives

$$\Sigma a^2 = Q_1\Sigma\, aa + Q_2\Sigma\, ba + \ldots Q_q\Sigma\, qa$$
$$= Q_1 \qquad \text{by (H) and (H′)} \qquad \text{(I)}$$

Let μ be the mean error of an observation of weight unity.
Let μ_{z_1} be the mean error of z_1.
Let p_{z_1} be the weight of z_1.

The mean errors of $m_1, m_2, \ldots m_n$ are the same as the mean errors of $M_1, M_2, \ldots M_n$ and each is accordingly equal to μ. Therefore from (G), by (69)

$$\mu_{z_1}{}^2 = a_1{}^2\mu^2 + a_2{}^2\mu^2 + \ldots a_n{}^2\mu^2$$
$$= \mu^2\Sigma\, a^2$$
$$= Q_1\mu^2 \text{ by (I)} \qquad \text{(J)}$$

But by (48) $\quad \mu_{z_1}{}^2 = \dfrac{\mu^2}{p_{z_1}}$

Comparing this with (J) we see at once that

$$Q_1 = \frac{1}{p_{z_1}} \qquad \text{(K)}$$

Therefore, for the First Method of computing the weights we have the following:—

77. **Rule I.** *In the normal equation for* z_1 *write* -1 *for the absolute term* Σam, *and in the other equations zero for each of the absolute terms* $\Sigma bm, \Sigma cm, \ldots \Sigma qm$. *The value of* z_1 *found from these equations, is the reciprocal of the weight of the value of* z_1 *obtained by the solution of the normal equations.*

To find the weights of $z_2, z_3, \ldots z_q$, *proceed in a similar way, forming a corresponding set of equations for each unknown.*

SECOND METHOD OF COMPUTING THE WEIGHTS.

78. Write equations (B) of paragraph 76 in the form

$$\begin{array}{l} z_1 \Sigma a^2 + z_2 \Sigma ab + \ldots z_q \Sigma aq + \Sigma am = A \\ z_1 \Sigma ab + z_2 \Sigma b^2 + \ldots z_q \Sigma bq + \Sigma bm = B \\ \quad \ldots \qquad \ldots \qquad \ldots \\ \quad \ldots \qquad \ldots \qquad \ldots \\ z_1 \Sigma aq + z_2 \Sigma bq + \ldots z_q \Sigma q^2 + \Sigma qm = Q \end{array} \qquad \text{(L)}$$

Then in the solution by the preceding method, equation (E) becomes

$$\begin{array}{l} z_1 + Q_1 \Sigma am + Q_2 \Sigma bm + \ldots Q_q \Sigma qm \\ \quad = Q_1 A + Q_2 B + \ldots Q_q Q \end{array} \qquad \text{(M)}$$

in which, as was proved in (K), Q_1 is the reciprocal of the weight of z_1. Whatever method of elimination is employed in the solution of the normal equations, the coefficient of A in the value of z_1 must necessarily be always the same. Hence we have

79. Rule II. *Write* $A, B, \ldots Q$ *instead of zero in the second members of the normal equations and carry out their solution in any convenient way. Then the most probable values of* $z_1, z_2, \ldots z$ *are given by those terms in the results which are independent of* $A, B, \ldots Q$.

The weight of z_1 *is the reciprocal of the coefficient of* A *in the value of* z_1. *The weight of* z_2 *is the reciprocal of the coefficient of* B *in the value of* z_2, *etc., etc.*

THIRD METHOD OF COMPUTING THE WEIGHTS.

80. From the second, third, ... equations of (L) find the values of $z_2, z_3, \ldots z_q$ in terms of z_1 and substitute in the first of (L) without reduction. Then the first of (L) becomes

$$Rz_1 = T + A + \text{terms in } B, C, \ldots Q$$

Where T is the sum of all the numerical quantities resulting from the substitutions. Dividing through by R,

$$z_1 = \frac{T}{R} + \frac{A}{R} + \text{terms in } B, C, \ldots Q \qquad \text{(N)}$$

in which $\frac{T}{R}$ is the most probable value of z_1, and, as was shown in deriving the second method,

$$\frac{A}{R} = Q_1 A$$

$$\frac{1}{R} = Q_1$$

$$\therefore \qquad R = p_{z_1} \qquad \text{(O)}$$

From this follows at once

81. Rule III. *Substitute in the normal equation for* z_1 *the values of* $z_2, z_3, \ldots z_q$ *in terms of* z_1 *as found from the remaining equations. Then before freeing of fractions or introducing any reduction factor, the coefficient of* z_1

in this equation is the weight of the value of z_1 *obtained in the solution.*

To find the weights of $z_2, z_3, \ldots z_q$, *proceed in a similar way with the normal equations for each of these unknowns.*

For the solution of an example by the three different methods see paragraph 84.

THE MEAN ERROR OF AN OBSERVATION.

82. The next step will be to derive μ, the mean error of an observation of weight unity. In the following demonstration the equations referred to by letters are those in paragraphs 76 to 81.

Let the real values of $z_1, z_2, \ldots z_q$ be

$$z_1 + x_1, \quad z_2 + x_2, \ldots \quad z_q + x_q$$

and substituting in (A) we have

$$\begin{array}{l} a_1(z_1 + x_1) + b_1(z_2 + x_2) + \ldots q_1(z_q + x_q) + m_1 = \Delta_1 \\ a_2(z_1 + x_1) + b_2(z_2 + x_2) + \ldots q_2(z_q + x_q) + m_2 = \Delta_2 \\ \ldots \qquad \ldots \qquad \ldots \qquad \qquad (P) \\ \ldots \qquad \ldots \qquad \ldots \\ a_n(z_1 + x_1) + b_n(z_2 + x_2) + \ldots q_n(z_q + x_q) + m_n = \Delta_n \end{array}$$

where $\Delta_1, \Delta_2, \ldots \Delta_n$ are the real errors of $M_1, M_2, \ldots M_n$, or of $m_1, m_2, \ldots m_n$.

Multiply the first of (P) by a_1, the second by $a_2, \ldots$ and add the results. This gives

$$\begin{array}{l} z_1 \Sigma a^2 + z_2 \Sigma ab + \ldots z_q \Sigma aq + \Sigma am \\ + x_1 \Sigma a^2 + x_2 \Sigma ab + \ldots x_q \Sigma aq \qquad = \Sigma a \Delta \end{array}$$

But by the first of (B) the first line in this equation is equal to zero, and therefore

$$x_1 \Sigma a^2 + x_2 \Sigma ab + \ldots x_q \Sigma aq - \Sigma a\Delta = 0$$

Also

$$\begin{aligned} x_1 \Sigma ab + x_2 \Sigma b^2 + \ldots x_q \Sigma bq - \Sigma b\Delta &= 0 \\ \ldots \quad \ldots \quad \ldots & \\ \ldots \quad \ldots \quad \ldots & \\ x_1 \Sigma aq + x_2 \Sigma bq + \ldots x_q \Sigma q^2 - \Sigma q\Delta &= 0 \end{aligned} \qquad \text{(Q)}$$

These being of the same form as the normal equations (B), the value of x_1 resulting from their solution will be of the same form as that of z_1 in the solution of those equations, with only the substitution of $-\Delta$ for m. From (G) we shall therefore have

$$x_1 - a_1\Delta_1 - a_2\Delta_2 - \ldots a_n\Delta_n = 0 \qquad \text{(R)}$$

Multiply the first of (P) by v_1, the second of (P) by v_2, . . . and add the results. Then

$$(z_1 + x_1)\Sigma av + (z_2 + x_2)\Sigma bv + \ldots (z_q + x_q)\Sigma qv + \Sigma mv = \Sigma \Delta v$$

and multiplying the first of (A) by a_1, the second by a_2, . . . and adding the results

$$\begin{aligned} \Sigma av &= z_1 \Sigma a^2 + z_2 \Sigma ab + \ldots z_q \Sigma aq + \Sigma am \\ &= 0 \qquad \text{by the first of (B).} \end{aligned}$$

Also

$$\begin{aligned} \Sigma bv &= 0 \\ &\ldots \\ \Sigma qv &= 0 \end{aligned}$$

Substituting these values in the above, we find that

$$\Sigma mv = \Sigma \Delta v \qquad \text{(S)}$$

Now multiply the first of (A) by v_1, the second by v_2, . . . and add the results. This shows that

$$z_1 \Sigma av + z_2 \Sigma bv + \ldots z_q \Sigma qv + \Sigma mv = \Sigma v^2$$

and as above,

$$\Sigma av = \Sigma bv \ldots = \Sigma qv = 0$$

Combining this result with (S) we have

$$\Sigma mv = \Sigma v^2 = \Sigma \Delta v \qquad \text{(T)}$$

Next multiply the first of (A) by Δ_1, the second by $\Delta_2, \ldots$, add the results and compare with (T). This gives

$$z_1 \Sigma a\Delta + z_2 \Sigma b\Delta + \ldots z_q \Sigma q\Delta + \Sigma m\Delta$$
$$= \Sigma \Delta v = \Sigma v^2 \qquad \text{(U)}$$

And finally, multiplying the first of (P) by Δ_1, the second by $\Delta_2, \ldots$ and adding the results, we have

$$z_1 \Sigma a\Delta + z_2 \Sigma b\Delta + \ldots z_q \Sigma q\Delta + \Sigma m\Delta$$
$$+ x_1 \Sigma a\Delta + x_2 \Sigma b\Delta + \ldots x_q \Sigma q\Delta = \Sigma \Delta^2$$

Therefore from (U)

$$\Sigma v^2 + x_1 \Sigma a\Delta + x_2 \Sigma b\Delta + \ldots x_q \Sigma q\Delta = \Sigma \Delta^2 \qquad \text{(V)}$$

$$\therefore \quad \mu^2 = \frac{\Sigma \Delta^2}{n} = \frac{\Sigma v^2}{n} + \frac{x_1 \Sigma a\Delta + \ldots x_q \Sigma q\Delta}{n} \qquad \text{(W)}$$

We must now find the mean values of the terms $x_1 \Sigma a\Delta$, $x_2 \Sigma b\Delta, \ldots x_q \Sigma q\Delta$. Expanding $\Sigma a\Delta$,

$$\Sigma a\Delta = a_1\Delta_1 + a_2\Delta_2 + \ldots a_n\Delta_n$$

from (R) $\quad x_1 = \alpha_1\Delta_1 + \alpha_2\Delta_2 + \ldots \alpha_n\Delta_n$

Multiplying,

$$x_1 \Sigma a\Delta = a_1\alpha_1\Delta_1^2 + a_2\alpha_2\Delta_2^2 + \ldots a_n\alpha_n\Delta_n^2$$
$$+ \text{ terms in } \Delta_1\Delta_2,\ \Delta_1\Delta_3, \ldots$$

In the most probable case these product terms vanish, and substituting for $\Delta_1^2, \Delta_2^2, \ldots \Delta_n^2$ the mean value μ^2, we have

$$x_1 \Sigma a\Delta = \mu^2 \Sigma aa$$

$$= \mu^2 \quad \text{by (H)}$$

Similarly

$$x_2 \Sigma b\Delta = \mu^2$$

$$\cdots$$

$$\cdots$$

$$x_q \Sigma q\Delta = \mu^2$$

From (W) then, q being the number of unknowns,

$$\mu^2 = \frac{\Sigma v^2}{n} + \frac{q\,\mu^2}{n}$$

$$\mu = \sqrt{\frac{\Sigma v^2}{n - q}} \tag{72}$$

$$\mu_z = \frac{\mu}{\sqrt{p_z}} = \sqrt{\frac{\Sigma v^2}{p_z(n - q)}} \tag{73}$$

$$r_z = .6745\,\mu_z = .6745\sqrt{\frac{\Sigma v^2}{p_z(n - q)}} \tag{74}$$

83. By a method similar to that used in paragraph 60, we may derive

$$a.d. = \frac{\Sigma v}{\sqrt{n(n - q)}} \tag{75}$$

$$a.d._z = \frac{\Sigma v}{\sqrt{p_z n(n - q)}} \tag{76}$$

84. *Example.* In illustration of the above processes take the example in paragraph 24, where we found for

OBSERVATION EQUATIONS

$$\left.\begin{array}{rrrrcl} z_1 & - z_2 & & - 1.7 & = & v_1 \\ & & z_3 & - 2.4 & = & v_2 \\ - z_1 & + z_2 & + z_3 & - 1.0 & = & v_3 \\ & z_2 & - z_3 & - 3.0 & = & v_4 \end{array}\right\} \quad (A)$$

and for

NORMAL EQUATIONS

$$\begin{array}{rrrrcl l} 2 z_1 & - 2 z_2 & - z_3 & - 0.7 & = & 0 & \text{(a)} \\ - 2 z_1 & + 3 z_2 & & - 2.3 & = & 0 & \text{(b)} \\ - z_1 & & + 3 z_3 & - 0.4 & = & 0 & \text{(c)} \end{array}$$

SOLUTION FOR THE WEIGHTS BY THE FIRST METHOD.

For finding the weight of z_1 the above normal equations would be written

$$\begin{array}{rrrrcl l} 2 z_1 & - 2 z_2 & - z_3 & - 1 & = & 0 & \text{(a')} \\ - 2 z_1 & + 3 z_2 & & & = & 0 & \text{(b')} \\ - z_1 & & + 3 z_3 & & = & 0 & \text{(c')} \end{array}$$

Solve for z_1

$$\begin{array}{l rrrrcl} 3 \times (a') & 6 z_1 & - 6 z_2 & - 3 z_3 & - 3 & = & 0 \\ (c') & - z_1 & & + 3 z_3 & & = & 0 \\ \hline & 5 z_1 & - 6 z_2 & & - 3 & = & 0 \\ 2 \times (b') & - 4 z_1 & + 6 z_2 & & & = & 0 \\ \hline & z_1 & & & - 3 & = & 0 \end{array}$$

$$z_1 = 3 \quad \therefore \quad p_{z_1} = \frac{1}{3} \qquad (d')$$

For finding the weight of z_2 the equations are

$$\begin{array}{rrrrcl l} 2 z_1 & - 2 z_2 & - z_3 & & = & 0 & \text{(a'')} \\ - 2 z_1 & + 3 z_2 & & - 1 & = & 0 & \text{(b'')} \\ - z_1 & & + 3 z_3 & & = & 0 & \text{(c'')} \end{array}$$

Solve for z_2

$$\begin{array}{lrcl}
3 \times (a'') & 6\,z_1 - 6\,z_2 - 3\,z_3 & = & 0 \\
(c'') & -\;z_1 \qquad\qquad + 3\,z_3 & = & 0 \\ \hline
 & 5\,z_1 - 6\,z_2 \qquad\qquad & = & 0
\end{array}$$

$$z_1 = \frac{6}{5}\,z_2$$

Substitute in (b'') $\quad z_2 = \dfrac{5}{3} \quad \therefore \quad p_{z_2} = \dfrac{3}{5} \qquad (d'')$

For finding the weight of z_3 the equations are

$$\begin{array}{rcll}
2\,z_1 - 2\,z_2 - \quad z_4 \qquad\quad & = & 0 & (a''') \\
-\,2\,z_1 + 3\,z_2 \qquad\qquad\qquad\quad & = & 0 & (b''') \\
-\quad z_1 \qquad\qquad + 3\,z_3 - 1 & = & 0 & (c''')
\end{array}$$

Solve for z_3

from (b''') $\quad 2\,z_2 = \dfrac{4}{3}\,z_1$

$$\begin{array}{lrcl}
\text{substitute in } (a''') & 2\,z_1 - 3\,z_3 \qquad & = & 0 \\
2 \times (c''') & -\,2\,z_1 + 6\,z_3 - 2 & = & 0 \\ \hline
 & 3\,z_3 - 2 & = & 0
\end{array}$$

$$z_3 = \frac{2}{3} \quad \therefore \quad p_{z_3} = \frac{3}{2} \qquad (d''')$$

SOLUTION BY THE SECOND METHOD.

The normal equations will now be modified so as to appear in the following form:–

$$\begin{array}{rcll}
2\,z_1 - 2\,z_2 - \quad z_3 - 0.7 & = & A & (a) \\
-\,2\,z_1 + 3\,z_2 \qquad\qquad - 2.3 & = & B & (b) \\
-\quad z_1 \qquad\qquad + 3\,z_3 - 0.4 & = & C & (c)
\end{array}$$

Solving,

$$
\begin{array}{lrl}
3 \times (a) & 6\,z_1 - 6\,z_2 - 3\,z_3 - 2.1 & = 3\,A \\
(c) & -\;z_1 \qquad\qquad + 3\,z_3 - 0.4 & = C \\
\hline
 & 5\,z_1 - 6\,z_2 - 2.5 & = 3\,A + C \\
 & -\,4\,z_1 + 6\,z_2 - 4.6 & = 2\,B \\
\hline
 & z_1 \qquad - 7.1 & = 3\,A + 2\,B + C \qquad (\delta)
\end{array}
$$

$$\therefore \qquad z_1 = 7.1 \quad \text{and} \quad p_{z_1} = \frac{1}{3} \qquad \text{(d)}$$

Substituting (δ) in (b)

$$3\,z_2 = 16.5 + 6\,A + 5\,B + 2\,C$$

$$\therefore \qquad z_2 = 5.5 \quad \text{and} \quad p_{z_2} = \frac{3}{5} \qquad \text{(e)}$$

Substituting (δ) in (c)

$$3\,z_3 = 7.5 + 3\,A + 2\,B + 2\,C$$

$$\therefore \qquad z_3 = 2.5 \quad \text{and} \quad p_{z_3} = \frac{3}{2} \qquad \text{(f)}$$

SOLUTION BY THE THIRD METHOD.

The normal equations are now taken in their original form.

$$
\begin{array}{rcr}
2\,z_1 - 2\,z_2 - z_3 - 0.7 = 0 & & \text{(a)} \\
-\,2\,z_1 + 3\,z_2 \qquad - 2.3 = 0 & & \text{(b)} \\
-\,z_1 \qquad + 3\,z_3 - 0.4 = 0 & & \text{(c)}
\end{array}
$$

To obtain z_1 and its weight we proceed as follows:

from (c) $$z_3 = \frac{z_1}{3} + \frac{.4}{3}$$

from (b) $$z_2 = \frac{2\,z_1}{3} + \frac{2.3}{3}$$

Substitute in (a)

$$2\,z_1 - \frac{4}{3}z_1 - \frac{4.6}{3} - \frac{z_1}{3} - \frac{.4}{3} - 0.7 = 0$$

Collecting terms, $$\frac{z_1}{3} - \frac{7.1}{3} = 0$$

$$\therefore \quad z_1 = 7.1 \quad \text{and} \quad p_{z_1} = \frac{1}{3} \qquad \text{(d)}$$

For z_2

3 × (a) + (c) $$5\,z_1 - 6\,z_2 - 2.5 = 0$$

$$\therefore \quad z_1 = \frac{6}{5}z_2 + 0.5$$

Substitute in (b) $$-\frac{12}{5}z_2 - 1.0 + 3\,z_2 - 2.3 = 0$$

Collecting terms, $$\frac{3}{5}z_2 - 3.3 = 0$$

$$\therefore \quad z_2 = 5.5 \quad \text{and} \quad p_{z_2} = \frac{3}{5} \qquad \text{(e)}$$

For z_3

3 × (a) + 2 × (b) $$2\,z_1 - 3\,z_3 - 6.7 = 0$$

$$\therefore \quad z_1 = \frac{3}{2}z_3 + \frac{6.7}{2}$$

Substitute in (c) $$-\frac{3}{2}z_3 - \frac{6.7}{2} + 3\,z_3 - 0.4 = 0$$

Collecting terms, $$\frac{3}{2} z_3 - \frac{7.5}{2} = 0$$

$$\therefore \quad z_3 = 2.5 \quad \text{and} \quad p_{z_3} = \frac{3}{2} \qquad \text{(f)}$$

It is evident that the three methods give identically the same results and that the work is about the same in each case.

COMPUTATION OF THE PRECISION MEASURES.

Substituting the values found for z_1, z_2 and z_3 in the observations equations (A), we have

$$\begin{aligned}
7.1 - 5.5 - 1.7 &= v_1 = -.1 & .01 &= v_1^2 \\
2.5 - 2.4 &= v_2 = +.1 & .01 &= v_2^2 \\
-7.1 + 5.5 + 2.5 - 1.0 &= v_3 = -.1 & .01 &= v_3^2 \\
5.5 - 2.5 - 3.0 &= v_4 = .0 & .00 &= v_4^2 \\
& & \overline{.03} &= \Sigma v^2
\end{aligned}$$

In this example $n = 4$, $q = 3$.

By (72) $$\mu = \sqrt{\frac{.03}{1}} = .17$$

By (74) $$r = .6745\mu = .12$$

By (73) $$\mu_{z_1} = \sqrt{\frac{.03}{\frac{1}{3}}} = .30 \qquad r_{z_1} = .20$$

$$\mu_{z_2} = \sqrt{\frac{.03}{\frac{3}{5}}} = .22 \qquad r_{z_2} = .15$$

$$\mu_{z_3} = \sqrt{\frac{.03}{\frac{3}{2}}} = .14 \qquad r_{z_3} = .10$$

By (75) $$a.d. = \frac{.3}{2} = .15$$

$$r = .8453\ a.d. = .13$$

85. Observations of Unequal Weights. If the observations are not all of the same weight the formulas and operations are merely modified in the usual manner and equations (72) and (75) take the more general form

$$\mu = \sqrt{\frac{\Sigma pv^2}{n - q}} \tag{77}$$

$$a.d. = \frac{\Sigma v\sqrt{p}}{\sqrt{n(n - q)}} \tag{78}$$

CONDITIONED OBSERVATIONS.

86. Suppose there are given n observations, n' conditions and q unknown quantities. Then by paragraph 33, the method of solution is to eliminate n' unknowns between the "Observation" and "Condition" equations, leaving $q - n'$ independent unknowns in the "Normal" equations. Consequently formula (77) now applies to this case and it would be written

$$\mu = \sqrt{\frac{\Sigma pv^2}{n - q + n'}} \tag{79}$$

also
$$\mu_z = \sqrt{\frac{\Sigma pv^2}{p_z(n - q + n')}} \tag{80}$$

The weights of the $q - n'$ unknown quantities can be found by any one of the three methods already given and then the mean error of each unknown may be computed by using formula (80). If the mean errors of the n' quantities that were first eliminated is wanted, their weights must be determined by eliminating a different set of n' quantities from the original observation equations and solving the necessary sets of equations for these weights. The first method of solution for the weights would perhaps be best here as the actual values of the unknowns have already been found.

87. *Example.* Given the

OBSERVATION EQUATIONS

$$\begin{array}{lllll} z_1 + z_2 & & - 3.0 = 0 & \text{weight } 1 & \\ z_1 & - z_3 & + 1.5 = 0 & \text{"} \quad 4 & \\ \quad\; z_2 & & - 2.2 = 0 & \text{"} \quad 3 & (A) \\ z_1 & + z_3 & - 3.4 = 0 & \text{"} \quad 2 & \end{array}$$

and the

CONDITION EQUATION

$$z_3 - z_2 - 0.5 = 0 \qquad (B)$$

To find the most probable values of z_1, z_2, and z_3, and also their mean errors.

Eliminating z_3 between (A) and (B) there remains

$$\begin{array}{ll} z_1 + z_2 - 3.0 = 0 & \text{weight } 1 \\ z_1 - z_2 + 1.0 = 0 & \text{"} \quad 4 \\ \quad\;\; z_2 - 2.2 = 0 & \text{"} \quad 3 \\ z_1 + z_2 - 2.9 = 0 & \text{"} \quad 2 \end{array}$$

From these we have the

NORMAL EQUATIONS

$$\left.\begin{array}{r} 7\, z_1 - \quad z_2 - \;\; 4.8 = 0 \\ - z_1 + 10\, z_2 - 19.4 = 0 \end{array}\right\} \qquad (C)$$

Solving,

$$\begin{array}{lll} & z_1 = 0.98 & p_{z_1} = 6.9 \\ & z_2 = 2.04 & p_{z_2} = 9.9 \qquad (D) \\ \text{from (B)} & z_3 = 2.54 & \end{array}$$

Now eliminating z_2 between (A) and (B), we find

$$\begin{aligned} z_1 + z_3 - 3.5 &= 0 \qquad \text{weight } 1 \\ z_1 - z_3 + 1.4 &= 0 \qquad \text{"} \quad 4 \\ z_3 - 2.7 &= 0 \qquad \text{"} \quad 3 \\ z_1 + z_3 - 3.4 &= 0 \qquad \text{"} \quad 2 \end{aligned} \tag{E}$$

From these we derive the new set of

NORMAL EQUATIONS

$$\begin{aligned} 7\,z_1 - z_3 - 4.3 &= 0 \\ -\,z_1 + 10\,z_3 - 24.4 &= 0 \end{aligned} \tag{F}$$

and solving for z_3

$$z_3 = 2.54 \qquad p_{z_3} = 9.9 \tag{G}$$

Substituting the values of z_1, z_2 and z_3 in equations (A), we have the residuals

v	v^2	pv^2
.02	.0004	.0004
.06	.0036	.0144
.16	.0256	.0768
.12	.0144	.0288
		.1204 $= \Sigma pv^2$

Since in this example $n = 4$, $n' = 1$, $q = 3$,

By (79) $$\mu = \sqrt{\frac{.1204}{2}} = .25$$

$$\therefore \qquad \mu_{z_1} = \frac{\mu}{\sqrt{6.9}} = .09$$

$$\mu_{z_2} = \frac{\mu}{\sqrt{9.9}} = .08$$

$$\mu_{z_3} = \frac{\mu}{\sqrt{9.9}} = .08$$

The first significant figure in the mean errors is so large that it is not worth while to retain the second place as usual.

If desired, the probable errors and average deviations can now be computed by the usual formulas.

88. In case the observations are made directly upon the values of several quantities subject to certain conditions, we have $n = q$, and equation (79) reduces to

$$\mu = \sqrt{\frac{\Sigma pv^2}{n'}} \tag{81}$$

from which the mean error of any observation may at once be computed from its weight.

89. *Example.* Taking the example in paragraph 34 on the measurement of the angles of a quadrilateral, we had for

OBSERVATION EQUATIONS

$$\begin{array}{lll} z_1 = 0 & \text{weight } 3 & \\ z_2 = 0 & \quad " \quad 2 & \\ z_3 = 0 & \quad " \quad 2 & \qquad (A) \\ z_4 = 0 & \quad " \quad 1 & \end{array}$$

for the

CONDITION EQUATION

$$z_1 + z_2 + z_3 + z_4 + 58 = 0 \tag{B}$$

and for the

NORMAL EQUATIONS

$$\begin{array}{l} 4\,z_1 + z_2 + z_3 + 58 = 0 \\ z_1 + 3\,z_2 + z_3 + 58 = 0 \qquad (C) \\ z_1 + z_2 + 3\,z_3 + 58 = 0 \end{array}$$

Solving these equations for the values of the unknown quantities and also for their weights, we have

$$\begin{array}{ll} z_1 = -\ 8.3 & p_{z_1} = 3.5 \\ z_2 = -\ 12.4 & p_{z_2} = 2.5 \\ z_3 = -\ 12.4 & p_{z_3} = 2.5 \end{array}$$

and forming a new set of normals containing z_4, we find, on solving for that quantity,

$$z_4 = -24.9 \qquad p_{z_4} = 1.75$$

These values of z_1, z_2, z_3, z_4 are also in this case the residuals of the observations, and therefore to compute the precision measures we have

v	v^2	pv^2
8.3	68.9	206.7
12.4	153.8	307.6
12.4	153.8	307.6
24.9	620.0	620.0
		1441.9 $= \Sigma pv^2$

$$n' = 1$$

By (81)

$$\mu = \sqrt{\frac{1441.9}{1}} = 38$$

$$\mu_{z_1} = \frac{\mu}{\sqrt{3.5}} = 20 \qquad r_{z_1} = 13$$

$$\mu_{z_2} = \mu_{z_3} = \frac{\mu}{\sqrt{2.5}} = 24 \qquad r_{z_2} = r_{z_3} = 16$$

$$\mu_{z_4} = \frac{\mu}{\sqrt{1.75}} = 29 \qquad r_{z_4} = 20$$

We shall accordingly write for the most probable values of the angles of the quadrilateral

A =	101°	13′	14″	±	13″
B =	93	49	5	±	16
C =	87	5	27	±	16
D =	77	52	15	±	20

In the original solution in paragraph 34 it is evident that the results were carried out to a greater number of places of significant figures than the character of the observations warranted.

CHAPTER V.

MISCELLANEOUS THEOREMS.

THE DISTRIBUTION OF ERRORS.

90. Having developed the processes for the adjustment of observations according to the Method of Least Squares, it will now be interesting to show how closely the distribution of errors found in actual practice corresponds to the theoretical distribution upon which our methods of solution are based.

By formula (36), the probability that the error of a single observation will be numerically less than a is

$$P = \frac{2h}{\sqrt{\pi}} \int_0^a e^{-h^2x^2} dx \tag{82}$$

Let $t = hx$, $\therefore\ dt = h\,dx$. Also when $x = 0$, $t = 0$; and when $x = a$, $t = ha = \rho \frac{a}{r}$. Substituting in (82),

$$P = \frac{2}{\sqrt{\pi}} \int_0^{\rho \frac{a}{r}} e^{-t^2} dt \tag{83}$$

Values of P for values of the argument $\frac{a}{r}$ are given in Table I. Also for any series of observations this quantity P will represent the fraction of the entire number which should have errors less than the amount a. Hence if P is multiplied by the whole number of observations the result will be the number of errors which should be less than the limit a.

91. *Example.* Forty measurements on the diameter of Saturn's ring were made by Bessel, with the following results:—

M	v	M	v	M	v	M	v
38″.91	−.40	39″.35	+.04	39″.41	+.10	39″.02	−.29
39 .32	+.01	39 .25	−.06	39 .40	+.09	39 .01	−.30
38 .93	−.38	39 .14	−.17	39 .36	+.05	38 .86	−.45
39 .31	.00	39 .47	+.16	39 .20	−.11	39 .51	+.20
39 .17	−.14	39 .29	−.02	39 .42	+.11	39 .21	−.10
39 .04	−.27	39 .32	+.01	39 .30	−.01	39 .17	−.14
39 .57	+.26	39 .40	+.09	39 .41	+.10	39 .60	+.29
39 .46	+.15	39 .33	+.02	39 .43	+.12	39 .54	+.23
39 .30	−.01	39 .28	−.03	39 .43	+.12	39 .45	+.14
39 .03	−.28	39 .62	+.31	39 .36	+.05	39 .72	+.41

From these the most probable value of the diameter is found to be

$$D = 39''.308 \pm 0''.022$$

the probable error of a single observation being $r = 0''.136$.

Compare the theoretical and actual distribution of errors

between	0″.00	and	0″.05
"	0 .05	"	0 .10
"	0 .10	"	0 .20
"	0 .20	"	0 .30
"	0 .30	"	0 .40
over	0 .40		

In the following table the first column gives the successive values of the limiting error a, the second column the values of $\frac{a}{r}$, and the third column the corresponding values of P.

The fourth column contains the differences between the successive values of P, and by multiplying each of these differences by 40, the number of observations, we have the quantities in column five, which are the numbers of errors that according to the theory should fall within the corresponding limits. Column six shows the actual number of residuals occurring between these limits.

a	$\frac{a}{r}$	P	d	n	n'
0.00	0.000	0.000			
			0.196	8	9
0.05	0.368	0.196			
			0.184	7	6
0.10	0.735	0.380			
			0.299	12	12
0.20	1.471	0.679			
			0.184	7	8
0.30	2.206	0.863			
			0.090	4	3
0.40	2.942	0.953			
			0.047	2	2
∞	∞	1.000			

This is a close agreement considering that the number of observations is not very large. Also the number of errors greater than the probable error should be equal to the number less than it. On counting the residuals we find twenty-one less than $0''.136$ and nineteen greater.

THE REJECTION OF OBSERVATIONS.

92. After a series of measurements have been made, it is frequently found that one or two of the observations differ widely from the others, and hence it becomes a matter of great importance to establish, if possible, some criterion by which we may determine whether such discordant observations should be rejected or not. We are not concerned here with the question of the detection of a mistake or constant error, which a consideration of the circumstances of the observations or of the instruments might reveal, but it is assumed that there is nothing whatever to guide us except the mere fact of the unusual size of the residuals of the observations under discussion. To reject an observation merely because it differs considerably from the others is entirely unjustifiable, while to retain it without any investigation is a neglect of the evidence furnished by the observations themselves.

The adoption of any rigid criterion based upon the magnitude of the residuals is perhaps more satisfactory from a mathematical standpoint than from that of a practical observer, and some of the latter are of the opinion that no observation should be rejected entirely, even the most widely discordant ones being given a certain weight. In this latter case, however, the Theory of Probability will furnish a guide as to the proper weights to assign to the different observations.

Of the various criteria that have been proposed, that developed by Pierce (see *Chauvenet*, page 558) is perhaps the most complete. The derivation and application of this criterion is, however, somewhat long and complicated, and for all ordinary cases the following simple methods will give practically as good results.

93. Criterion for the Rejection of a Single Doubtful Observation. It was shown in (83) that in a series of n observations the number of errors numerically less than a should be nP, and therefore the number of errors greater than a should be

$$n - nP = n(1 - P) \tag{84}$$

If the value of the expression in (84) is less than one-half, the occurrence of an error of magnitude a will have a greater probability against it than for it, and hence the observation corresponding may be rejected.

Accordingly the limit of rejection, a, of a single doubtful observation is obtained from the equation

$$n(1 - P) = \frac{1}{2}$$

or

$$P = \frac{2n - 1}{2n} \tag{85}$$

94. *Example.* Fifteen observations on the value of an angle are made. Ought any of the observations to be rejected?

M	v	v^2	v'	v'^2	v''	v''^2
2° 23′.90	− .30	.090	− .41	.168	− .33	.109
23 .76	− .44	.194	− .55	.303	− .47	.221
25 .21	+ 1.01	1.020	+ .90	.810		
24 .68	+ .48	.230	+ .37	.137	+ .45	.203
23 .96	− .24	.058	− .35	.123	− .27	.073
24 .26	+ .06	.004	− .05	.003	+ .03	.001
24 .83	+ .63	.397	+ .52	.270	+ .60	.360
24 .07	− .13	.017	− .24	.058	− .16	.026
23 .98	− .22	.048	− .33	.109	− .25	.063
24 .14	− .06	.004	− .17	.029	− .09	.008
24 .40	+ .20	.040	+ .09	.008	+ .17	.029
24 .38	+ .18	.032	+ .07	.005	+ .15	.023
24 .59	+ .39	.152	− .28	.078	+ .36	.130
24 .10	− .10	.010	+ .21	.044	− .13	.017
22 .80	− 1.40	1.960				
2° 24 .20 M_0		4.256 Σv^2		2.145 $\Sigma v'^2$		1.263 $\Sigma v''^2$

Using all the observations we find

$$M_0 = 2^\circ\ 24'.20 \qquad r = .6745\sqrt{\frac{4.256}{14}} = .37$$

By (85), $$P = \frac{29}{30} = .967$$

By Table I, $$\frac{a}{r} = 3.17 \quad \therefore \quad a = 1.17$$

As the residual -1.40 is larger than a, we reject the last observation.

From the remaining observations we now compute a new mean value and a new set of residuals. And we find

$$M'_0 = 2^\circ\ 24'.31 \qquad r' = .6745\sqrt{\frac{2.145}{13}} = .27$$

By (85), $$P = \frac{27}{28} = .964$$

By Table I, $$\frac{a}{r} = 3.11 \quad \therefore \quad a = .84$$

The third observation may accordingly be rejected.

From the thirteen observations that remain we find

$$M''_0 = 2^\circ\ 24'.23 \qquad r'' = .6745\sqrt{\frac{1.263}{12}} = .22$$

$$P = \frac{25}{26} = .962$$

$$\frac{a}{r} = 3.08 \quad \therefore \quad a = .68$$

Therefore no more observations are to be rejected.

95. The Huge Error. In cases where the number of observations is not unusually large, a simple and safe criterion for the rejection of a doubtful observation is found in the use of the "Huge Error."

This is an error of such a magnitude that 999 out of every 1000 errors are less than it and only 1 as large as or greater than it.

Therefore the probability that the error of any given observation will be less than the "Huge Error" is .999, and from Table I,

$$\text{when} \qquad P = .999, \qquad \frac{a}{r} = 4.9$$

$$\therefore \qquad a = \textit{Huge Error} = \mathbf{4.9}\, r$$
$$= \mathbf{3.3}\, \mu \qquad (86)$$
$$= \mathbf{4.1}\, a.d.$$

Then in any limited series of observations, if an error greater than the huge error is found, we should reject the observation corresponding.

See also, *Holman*, page 30; *Wright*, page 131.

CONSTANT ERRORS.

96. Throughout our discussion of the methods of adjusting observations so as to obtain from them the most probable values of the unknown quantities, all constant errors are supposed to have been eliminated before the Method of Least Squares is applied in deducing the results.

If this is not done, and each observation is subject to the same constant error, the final result will be affected by an equal amount, and in short, the Method of Least Squares is not capable of removing or reducing the effect of errors of this kind. All that is accomplished by the use of the method is to reduce to a minimum the effect of the Accidental Errors.

Hence it will be seen that although by increasing the number of observations of a given kind we may increase the precision, that is, reduce the probable error, of our final result as much as we choose, yet we do not in this way necessarily increase the accuracy of the determination.

But if the unknowns can be determined in several ways, or under a variety of different circumstances, with various instruments, or by different observers, then it is most probable that the constant errors of the different sets of measurements will be grouped about the true values of the unknowns according to the exponential law of error. Accordingly a combination of such observations will enable us to increase not only the precision, but also the accuracy of the final result, the constant errors of the different sets tending to cancel each other in the same way that the accidental errors of a single set do.

It is for this reason that determinations of a quantity from observations made in a variety of ways are more valuable than those obtained merely from different sets of measurements of the same kind.

97. The probability of the existence of a constant error may often be expressed in the following manner.

Example. A standard 100 ohm coil is compared with a Wheatstone's bridge and the mean result found to be 100.90 ± 0.20. To find the probability that there is an error in the bridge between $+ 0.30$ and $+ 1.50$ ohms.

Suppose the result 100.90 ± 0.20 is treated as a single observation, and we find by an application of (83) the probability that the error of this observation is numerically less than 0.60 ohms.

$$\text{Here} \qquad \frac{a}{r} = \frac{.60}{.20} = 3.00 \qquad \therefore \qquad P = .957$$

Hence, as far as is shown by the observations, the probability that 100.90 ohms is within 0.60 ohms of the true value is .957. But since it is known that the true resistance is 100 ohms, it follows that there is the same probability that there is a constant error in the bridge between + 0.30 and + 1.50 ohms.

98. Combination of Determinations having Different Constant Errors. In case two or more determinations of a quantity, together with their probable errors, are obtained, the method of combining them so as to secure the best final result was considered in paragraph 56, and in Example C, paragraph 57. But it was there assumed that all the results were subject to the same constant errors, while if this is not true the probable errors of the separate determinations bear no relation to their weights, and accordingly in such cases another process must be adopted.

To determine whether the different measurements may fairly be considered to have the same constant errors we may proceed as follows: —

Let the determinations of the quantity M be

$$M_1 \pm r_1 \qquad \text{(a)}$$

$$M_2 \pm r_2 \qquad \text{(b)}$$

and let the difference between these results be

$$d = M_1 - M_2 \qquad \text{(c)}$$

Then the probable error of d is by (66),

$$R = \sqrt{r_1^2 + r_2^2} \qquad \text{(d)}$$

If d is of such a magnitude that an accidental error as great as it may reasonably be expected, we may assume that the constant errors of (a) and (b) are the same, and proceed as in Example C, paragraph 57.

But if the probability of making two determinations which differ by the amount d is very small, we had best consider M_1 and M_2 to have the same weight, provided there are no special reasons for regarding one better than the other. The final value of M will then be the arithmetical mean of M_1 and M_2, and its probable error will be found by (53).

99. *Example A.* An angle is measured by a theodolite and by a transit with results

$$\begin{array}{ll} \text{By Theodolite,} & 24^\circ\ 13'\ 36''.0 \pm 3''.1 \\ \text{By Transit,} & 24^\circ\ 13'\ 24'' \pm 14'' \end{array}$$

What is the most probable value of the angle and its probable error?

Referring to the preceding paragraph,

$$r_1 = 3.1, \qquad r_2 = 14, \qquad d = 12,$$

and the probable error of d is

$$R = \sqrt{3.1^2 + 14^2} = 14$$

Then from Table I the probability that the accidental error of a determination will be at least as large as 12 is found from

$$\frac{a}{r} = \frac{12}{14} = .86 \quad \therefore \quad 1 - P = .57$$

That is, there is more than an even chance that two such determinations of the angle will differ by as much as 12.

Hence it is fair to assume that the two determinations are not affected by constant errors of different magnitudes, and they would be combined as in Example C, paragraph 57.

Example B. Suppose the zenith distance, M, of a star, observed at two different culminations, is found to be

$$\begin{array}{l} M_1 = 14^\circ\ 53'\ 12''.10 \pm 0''.30 \\ M_2 = 14^\circ\ 53'\ 14''.30 \pm 0''.50 \end{array}$$

What is the best final value?

Here $\quad d = 2.2, \qquad R = \sqrt{.09 + .25} = .58$

and for $\quad \dfrac{a}{r} = \dfrac{2.2}{.58} = 3.8, \qquad 1 - P = .01$

Therefore the chance that the difference in the two determinations, due to accidental errors, will be as large as 2.2 is only one in a hundred. It is to be concluded then that the constant errors of observations at the two culminations differ by about 2.2, and as there is nothing to show that one measurement is more accurate than the other we will give them both the same weight and take the mean. Then the best value for the zenith distance is

$$M_0 = 14° \ 53' \ 13''.20 \pm 0''.74$$

For $$M_0 = 14° \ 53' + \frac{12''.10 + 14''.30}{2}$$

$$= 14° \ 53' \ 13''.20$$

By (53) $$r_0 = .6745 \sqrt{\frac{2.42}{2 \times 1}} = .74$$

For a more extended treatment of this subject see *Johnson*, "The Theory of Errors and Method of Least Squares," chap. vii.

THE WEIGHTING OF OBSERVATIONS.

100. In case the relative worth of observations is not settled by methods already discussed, the proper weight to assign to each quantity in the final adjustment can only be determined from a full knowledge of all the circumstances of the measurements. Even then considerable experience in the particular work in hand is required before the best values for these weights can be assigned. The weight given to a quantity should never be considered final, but always subject to revision whenever new information with regard to the quantity is obtained. Thus an observation which at first is supposed to deserve a high degree of confidence is often found on later investigation to possess very little value, and vice versa.

See also *Wright*, page 118.

OTHER LAWS OF ERROR.

101. Although in the great majority of cases the distribution of errors follows the exponential law thus far considered, there are a few special cases in which some of the suppositions made in deriving that law do not hold, and hence for the adjustment of such observations the corresponding special laws of error must be determined.

For instance, in applying the exponential law we assume a large number of observations, that each observation is subject to the same law of error, that small errors are more likely to occur than large ones, and that positive and negative errors are equally probable. Now it is easy to conceive of cases where only positive errors can occur, or where the probability of the occurrence of a small error may not be greater than that of a larger one, etc. If we can determine the different sources of error in any case and the relative effect of each upon the quantity sought, we shall arrive at the law of error for that particular set of observations. The case of most common occurrence is the following.

102. Suppose all errors between the limits a and $-a$ are equally probable, and that there are no errors beyond these limits. Then if $y = \phi(x)$ is the equation of the Curve of Error, and its area is represented as in paragraph 18, we have

$$\int_{-a}^{a} \phi(x)\, dx = 1 \qquad \text{(a)}$$

or

$$2\,\phi(x) \int_{0}^{a} dx = 1 \qquad \text{(b)}$$

since by the supposition made $\phi(x)$ must be a constant.

Integrating and solving for $\phi(x)$, we have

$$y = \phi(x) = \frac{1}{2a}. \qquad (87)$$

To find the Mean Error we have by definition as in paragraph 52

$$\mu^2 = \int_{-a}^{a} x^2 \phi(x)\, dx$$

$$= \frac{1}{a}\int_0^a x^2\, dx$$

$$= \frac{a^2}{3}$$

$$\therefore \qquad \mu = \frac{a}{\sqrt{3}} \tag{88}$$

The Probable Error is derived from the equation

$$\int_{-r}^{r} \phi(x)\, dx = \frac{1}{2}$$

$$\frac{1}{a}\int_0^r dx = \frac{1}{2}$$

$$\therefore \qquad r = \frac{a}{2} \tag{89}$$

Finally, for the Average Deviation we have

$$a.d. = \int_{-a}^{a} x\, \phi(x)\, dx$$

$$= \frac{1}{a}\int_0^a x\, dx$$

$$\therefore \qquad a.d. = \frac{a}{2} \tag{90}$$

And the Curve of Error has the form

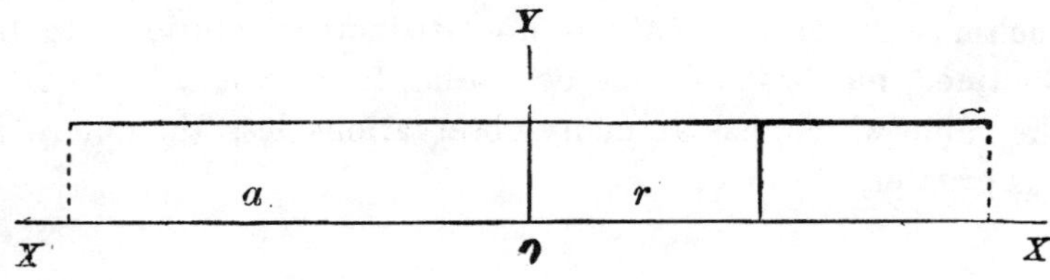

That the average deviation and probable error are in this case equal to one half of a may also be seen from the definition of these quantities.

Example. In taking a logarithm from a four place table, what is the probable error of the mantissa?

In this case the maximum error is .00005, and all errors between $-.00005$ and $.00005$ are equally probable. Therefore

$$r = .000025$$

103. The only other special case of common occurrence is that in which the error of a quantity is due to two sources, each of which can with the same probability assume all values between a and $-a$. Here it may be shown that the curve of error consists of two straight lines whose equations are

$$y = \frac{2a - x}{4a^2} \quad \text{and} \quad y = \frac{2a + x}{4a^2} \qquad (91)$$

Also $$\mu^2 = \frac{2}{3}a^2, \qquad r = (2 - \sqrt{2})\,a \qquad (92)$$

For a more extended discussion of special laws of error, see *Wright*, paragraphs 31 to 39. See also Example 151.

104. Contradictory Observations. Suppose three observations of a quantity give results 55, 56, 91. It is obvious that to take the arithmetical mean as the most probable value in this case would be contrary to the evidence furnished by the measurements, while in so small a number of observations it would not be allowable to reject the value 91 entirely. With such a series of observations no satisfactory solution can be obtained, but probably the best thing to do would be to take the value which has as many observations less than it as it has greater, or 56.

CHAPTER VI.

GAUSS'S METHOD OF SUBSTITUTION.

105. The most laborious part of the application of the Method of Least Squares to the adjustment of observations consists in the formation and solution of the "Normal Equations," and this labor increases enormously with increase in the number of observations, of unknowns and of conditions. It is not at all unusual to find that the adjustment of a single set of observations takes several weeks, even with all the aid that can be obtained from tables of logarithms, of squares, of products, and of reciprocals, and also from the use of calculating machines. In general the computations can be performed more rapidly and with less fatigue by using a machine or a table of products than by using logarithms, but a combination of methods is often desirable.

106. In any case however it is of the utmost importance that the formation and solution of the normal equations should be effected in a systematic way, and that as far as possible checks on the numerical work be carried along in the computations. By a slight amount of additional work checks upon the results at successive stages of the solution may be obtained by means indicated in the following demonstrations, while the most satisfactory form for the solution of the normal equations is given by the "Method of Substitution" proposed by *Gauss*.

This method will now be explained; and it is to be observed that it is customary in this subject to enclose a quantity in brackets when the sum of a number of quantities of the same kind is to be denoted. Thus $[ab]$ means the same as Σab.

For the sake of simplicity in demonstration it will be assumed that the observations are all reduced to weight unity.

107. Checks on the Formation of the Normal Equations. If, as in paragraph 76, we take for

OBSERVATION EQUATIONS

$$\left.\begin{array}{l} a_1z_1 + b_1z_2 + \dots q_1z_q + m_1 = v_1 \\ a_2z_1 + b_2z_2 + \dots q_2z_q + m_2 = v_2 \\ \dots \quad \dots \quad \dots \\ \dots \quad \dots \quad \dots \\ a_nz_1 + b_nz_2 + \dots q_nz_q + m_n = v_n \end{array}\right\} \qquad (A)$$

we shall have for

NORMAL EQUATIONS

$$\left.\begin{array}{l} [aa]\,z_1 + [ab]\,z_2 + \dots [aq]\,z_q + [am] = 0 \\ [ab]\,z_1 + [bb]\,z_2 + \dots [bq]\,z_q + [bm] = 0 \\ \dots \quad \dots \quad \dots \\ \dots \quad \dots \quad \dots \\ [aq]\,z_1 + [bq]\,z_2 + \dots [qq]\,z_q + [qm] = 0 \end{array}\right\} \qquad (B)$$

Let

$$\left.\begin{array}{l} a_1 + b_1 + \dots q_1 + m_1 = s_1 \\ a_2 + b_2 + \dots q_2 + m_2 = s_2 \\ \dots \quad \dots \quad \dots \\ \dots \quad \dots \quad \dots \\ a_n + b_n + \dots q_n + m_n = s_n \end{array}\right\} \qquad (C)$$

$$\therefore \quad [a] + [b] + \dots [q] + [m] = [s] \qquad (D)$$

Multiplying the first of (C) by m_1, the second by $m_2, \dots$ and adding, there results

$$[am] + [bm] + \dots [qm] + [mm] = [sm] \qquad (93)$$

Next multiplying each of equations (C) by its a and adding, and then each by its b and adding, etc., we have

$$
\begin{aligned}
&[aa] + [ab] + \ldots [aq] + [am] = [as] \\
&[ab] + [bb] + \ldots [bq] + [bm] = [bs] \\
&\quad \ldots \qquad \ldots \qquad \ldots \\
&\quad \ldots \qquad \ldots \qquad \ldots \\
&[aq] + [bq] + \ldots [qq] + [qm] = [qs]
\end{aligned}
\tag{94}
$$

Equation (93) will be satisfied if the absolute terms in the normal equations are correct, and equations (94) when the coefficients of the unknown quantities are correct. These check the formation of the normal equations.

108. The Reduced Normal Equations and the Elimination Equations.

The value of z_1 in terms of the remaining unknowns, derived from the first of equations (B), is

$$
z_1 = -\frac{[ab]}{[aa]} z_2 - \frac{[ac]}{[aa]} z_3 - \ldots \frac{[am]}{[aa]} \tag{E}
$$

Substituting this in the remaining $n - 1$ equations, they become

$$
\begin{aligned}
&\left\{[bb] - [ab]\frac{[ab]}{[aa]}\right\} z_2 + \ldots \left\{[bm] - [ab]\frac{[am]}{[aa]}\right\} = 0 \\
&\left\{[bc] - [ac]\frac{[ab]}{[aa]}\right\} z_2 + \ldots \left\{[cm] - [ac]\frac{[am]}{[aa]}\right\} = 0 \\
&\quad \ldots \qquad \ldots \qquad \ldots \\
&\quad \ldots \qquad \ldots \qquad \ldots \\
&\left\{[bq] - [aq]\frac{[ab]}{[aa]}\right\} z_2 + \ldots \left\{[qm] - [aq]\frac{[am]}{[aa]}\right\} = 0
\end{aligned}
$$

And letting

$$
\begin{aligned}
[bb, 1] &= [bb] - \frac{[ab][ab]}{[aa]} \\
[bc, 1] &= [bc] - \frac{[ab][ac]}{[aa]} \\
\ldots \quad &\ldots \quad \ldots \\
[qm, 1] &= [qm] - \frac{[aq][am]}{[aa]}
\end{aligned}
\tag{F}
$$

the above equations take the following form, which, being the same as that of the original normal equations, they are called the

FIRST REDUCED NORMAL EQUATIONS.

$$
\begin{array}{l}
[bb, 1]\, z_2 + [bc, 1]\, z_3 + \dots [bq, 1]\, z_q + [bm, 1] = 0 \\
[bc, 1]\, z_2 + [cc, 1]\, z_3 + \dots [cq, 1]\, z_q + [cm, 1] = 0 \\
\quad \dots \qquad \dots \qquad \dots \\
\quad \dots \qquad \dots \qquad \dots \\
[bq, 1]\, z_2 + [cq, 1]\, z_3 + \dots [qq, 1]\, z_q + [qm, 1] = 0
\end{array} \qquad (G)
$$

An inspection of equations (F) will render it easy to form a rule for writing out any one of them.

Now by means of the first of equations (G), eliminating z_2 from each of the others in the same way that z_1 was eliminated from the normal equations, there results the

SECOND REDUCED NORMAL EQUATIONS

$$
\begin{array}{l}
[cc, 2]\, z_3 + \dots [cq, 2]\, z_q + [cm, 2] = 0 \\
\quad \dots \qquad \dots \qquad \dots \\
\quad \dots \qquad \dots \qquad \dots \\
[cq, 2]\, z_3 + \dots [qq, 2]\, z_q + [qm, 2] = 0
\end{array} \qquad (H)
$$

In which

$$
\begin{array}{l}
[cc, 2] = [cc, 1] - \dfrac{[bc, 1]\,[bc, 1]}{[bb, 1]} \\
\quad \dots \qquad \dots \qquad \dots \\
\quad \dots \qquad \dots \qquad \dots \\
[qm, 2] = [qm, 1] - \dfrac{[bq, 1]\,[bm, 1]}{[bb, 1]}
\end{array} \qquad (I)
$$

Continuing this process we shall finally arrive at the single equation

$$
[qq, q-1]\, z_q + [qm, q-1] = 0 \qquad (J)
$$

from which the value of z_q is determined.

The value of z_{q-1} will then be obtained by substituting the numerical value of z_q in the first of the preceding set of equations, and so on, until finally z_1 is obtained from the first of the original normal equations. The equations from which the unknowns are actually determined are then the following, called the

ELIMINATION EQUATIONS.

$$\begin{array}{rl}
[aa]z_1 + [ab]z_2 + \cdots \quad [aq]z_q + [am] = 0 & \\
[bb,1]z_2 + \cdots [bq,1]z_q + [bm,1] = 0 & \\
\cdots \qquad \cdots \qquad \cdots & (95) \\
\cdots \qquad \cdots & \\
[qq, q-1]z_q + [qm, q-1] = 0 &
\end{array}$$

It may be seen from the rule in paragraph 81 that $[qq, q-1]$ is the weight of z_q, and the weight of any unknown might be found at the same time as its value by making it the last in the order of elimination, but except in special cases the weights had best be obtained by the general process of paragraph 115.

109. Check on the Solution of the Normal Equations. Multiplying the first of the Observation Equations (A) by m_1, the second by $m_2, \ldots$ and adding the results, we have

$$[mv] = [am]z_1 + [bm]z_2 + \cdots [qm]z_q + [mm]$$

But in equation (T), paragraph 82, it was shown that $[mv] = [vv]$. Therefore

$$[vv] = [am]z_1 + [bm]z_2 + \cdots [qm]z_q + [mm]$$

Substituting in this the value of z_1 from the first of (95), we get the result

$$[vv] = [bm,1]z_2 + [cm,1]z_3 + \cdots [qm,1]z_q + [mm,1]$$

in which

$$
\left.\begin{array}{lcl}
[bm, 1] & = & [bm] - \dfrac{[ab]\,[am]}{[aa]} \\
\cdots & \cdots & \cdots \\
\cdots & \cdots & \cdots \\
[mm, 1] & = & [mm] - \dfrac{[am]\,[am]}{[aa]}
\end{array}\right\} \qquad \text{(K)}
$$

being similar in form to equations (F).

Next eliminating z_2 in a like manner, we get

$$[vv] = [cm, 2]\, z_3 + \ldots [qm, 2]\, z_q + [mm, 2]$$

and continuing this process it finally appears that

$$[vv] = [mm, q] \qquad [96]$$

110. Arrangement of the Computations. In computing the coefficients that appear in the "Auxiliary" or "Reduced Normal Equations" it is most convenient to arrange the work in tabular form. The arrangement of the solution will be illustrated for an example containing four unknowns but it will be evident that the process can be extended to cover any case.

Let the Observation and Normal Equations be represented by equations (A) and (B), there being only four unknowns $z_1, z_2, z_3, z_4,$ and arrange a table as on the next page. In this scheme of solution the upper lines of the rows in the first compartment contain all the quantities that appear in the Normal Equations, together with $[mm]$ and the quantities in the column headed s which are used in checking the results in accordance with equations (93) and (94). The other compartments contain the corresponding quantities for the Reduced Normal Equations, and the first line in each compartment gives the coefficients in the Elimination Equations.

SCHEME A.—SOLUTION OF THE NORMAL EQUATIONS.

	a	b	c	d	m	s
a	[aa] log [aa]	[ab] log [ab]	[ac] log [ac]	[ad] log [ad]	[am] log [am]	[as] log [as]
b	log A_b	[bb] A_b [ab] log A_b [ab]	[bc] A_b [ac] log A_b [ac]	[bd] A_b [ad] log A_b [ad]	[bm] A_b [am] log A_b [am]	[bs] A_b [as] log A_b [as]
c	log A_c		[cc] A_c [ac] log A_c [ac]	[cd] A_c [ad] log A_c [ad]	[cm] A_c [am] log A_c [am]	[cs] A_c [as] log A_c [as]
d	log A_d			[dd] A_d [ad] log A_d [ad]	[dm] A_d [am] log A_d [am]	[ds] A_d [as] log A_d [as]
m	log A_m				[mm] A_m [am] log A_m [am]	[ms] A_m [as] log A_m [as]
b		[bb, 1] log [bb, 1]	[bc, 1] log [bc, 1]	[bd, 1] log [bd, 1]	[bm, 1] log [bm, 1]	[bs, 1] log [bs, 1]
c	log B_c		[cc, 1] B_c [bc, 1] log B_c [bc, 1]	[cd, 1] B_c [bd, 1] log B_c [bd, 1]	[cm, 1] B_c [bm, 1] log B_c [bm, 1]	[cs, 1] B_c [bs, 1] log B_c [bs, 1]
d	log B_d			[dd, 1] B_d [bd, 1] log B_d [bd, 1]	[dm, 1] B_d [bm, 1] log B_d [bm, 1]	[ds, 1] B_d [bs, 1] log B_d [bs, 1]
m	log B_m				[mm, 1] B_m [bm, 1] log B_m [bm, 1]	[ms, 1] B_m [bs, 1] log B_m [bs, 1]
c			[cc, 2] log [cc, 2]	[cd, 2] log [cd, 2]	[cm, 2] log [cm, 2]	[cs, 2] log [cs, 2]
d	log C_d			[dd, 2] C_d [cd, 2] log C_d [cd, 2]	[dm, 2] C_d [cm, 2] log C_d [cm, 2]	[ds, 2] C_d [cs, 2] log C_d [cs, 2]
m	log C_m				[mm, 2] C_m [cm, 2] log C_m [cm, 2]	[ms, 2] C_m [cs, 2] log C_m [cs, 2]
d				[dd, 3] log [dd, 3]	[dm, 3] log [dm, 3]	[ds, 3] log [ds, 3]
m	log D_m = log − z_4				[mm, 3] D_m [dm, 3] log D_m [dm, 3]	[ms, 3] D_m [ds, 3] log D_m [ds, 3]
m		z_4 =		[vv] =	[mm, 4]	[ms, 4]

The logarithms of the quantities in the first row of each compartment are also written in, and from these by proper subtractions are obtained the logarithms in the margin, where

$$
\begin{aligned}
A_b &= \frac{[ab]}{[aa]}, \quad A_c = \frac{[ac]}{[aa]}, \ \ldots \ A_m = \frac{[am]}{[aa]} \\
& \qquad\qquad B_c = \frac{[bc, 1]}{[bb, 1]}, \ \ldots \ B_m = \frac{[bm, 1]}{[bb, 1]} \\
& \qquad\qquad\qquad \ldots \qquad \ldots \\
& \qquad\qquad\qquad \ldots \qquad \ldots \\
& \qquad\qquad - z_4 = D_m = \frac{[dm, 3]}{[dd, 3]}
\end{aligned}
\qquad \text{(L)}
$$

Now in each compartment adding the logarithms at the margin to each of the logarithms in the first row of that compartment we obtain the corresponding logarithms written in the other rows. The numbers represented by these logarithms are next written above them, and if each of these quantities is then subtracted from the one above it the result will be the corresponding quantity in the compartment below. Some of the squares in each compartment are left vacant as the quantities belonging to them have already appeared above.

111. Application of Checks. Also, by (93) and (94), in the first compartment the quantities in the first lines of the last column should be equal to the sum of all the quantities in the first lines of the corresponding rows plus the quantities similarly situated above the first terms of the rows. Similar checks will apply in each compartment; for if from the second of equations (94) we subtract the product of the first equation multiplied by A_b we have

$$[bb, 1] + [bc, 1] + \ldots [bm, 1] = [bs, 1] \quad (97)$$

In the same manner we may show that a corresponding check holds throughout, so that finally we shall have

$$[mm, 4] = [ms, 4] \quad \textbf{(98)}$$

The last compartment of the table is added to give this final check and the value of $[vv]$ in accordance with equation (96).

If the multiplications and divisions are simple or if a table of squares or products or a computing machine is used the logarithms will of course be omitted from the scheme of solution.

112. *Example.* In order to illustrate the systematic formation of the coefficients that appear in the Normal Equations as well as the solution of the latter by the above method we will take the

OBSERVATION EQUATIONS

$$-z_1 + z_2 + z_3 + z_4 = 0.1 \qquad z_1 - z_2 \qquad - 2z_4 = 0.3$$
$$z_1 + z_2 - z_3 - z_4 = 0.6 \qquad z_2 - z_3 + z_4 = -0.1$$
$$z_1 - 2z_2 - z_3 + z_4 = 0.1 \qquad z_1 \qquad - z_3 \qquad = 0.4$$

First form a table containing the coefficients in these equations and also the sums s. As a first check the sum of the quantities in column s should be equal to the sum of all the quantities in all the other columns.

I. COEFFICIENTS IN THE OBSERVATION EQUATIONS.

No.	*a*	*b*	*c*	*d*	*m*	*s*
1	− 1	1	1	1	− .1	1.9
2	1	1	− 1	− 1	− .6	− .6
3	1	− 2	− 1	1	− .1	− 1.1
4	1	− 1	0	− 2	− .3	− 2.3
5	0	1	− 1	1	.1	1.1
6	1	0	− 1	0	− .4	− .4
Sum.	3	0	− 3	0	− 1.4	− 1.4

From these we now compute the coefficients in the Normal Equations (B), and also the necessary quantities for the check equations (93) and (94).

II. Coefficients in the Normal Equations.

No.	*aa*	*ab*	*ac*	*ad*	*am*	*as*	*bb*	*bc*	*bd*	*bm*
1	1	−1	−1	−1	.1	−1.9	1	1	1	−.1
2	1	1	−1	−1	−.6	−.6	1	−1	−1	−.6
3	1	−2	−1	1	−.1	−1.1	4	2	−2	.2
4	1	−1	0	−2	−.3	−2.3	1	0	2	.3
5	0	0	0	0	.0	.0	1	−1	1	.1
6	1	0	−1	0	−.4	−.4	0	0	0	.0
Sum	5	−3	−4	−3	−1.3	−6.3	8	1	1	−.1

No.	*bs*	*cc*	*cd*	*cm*	*cs*	*dd*	*dm*	*ds*	*mm*	*ms*
1	1.9	1	1	−.1	1.9	1	−.1	1.9	.01	−.19
2	−.6	1	1	.6	.6	1	.6	.6	.36	.36
3	2.2	1	−1	.1	1.1	1	−.1	−1.1	.01	.11
4	2.3	0	0	.0	.0	4	.6	4.6	.09	.69
5	1.1	1	−1	−.1	−1.1	1	.1	1.1	.01	.11
6	.0	1	0	.4	.4	0	.0	.0	.16	.16
Sum	6.9	5	0	.9	2.9	8	1.1	7.1	.64	1.24

If the coefficients in the Observation Equations are large, so that it becomes convenient to use logarithms, other tables corresponding to I and II would be formed to contain these logarithms.

Substituting these quantities now in the general tabular scheme of paragraph 110 we have the results on the following page. The work of the first compartment is performed without the use of logarithms, as the numbers are simple. The quantities in the last column should all be zero according to the check equations, what small differences there are being due to the rejection of figures beyond the third place in the decimals. The decimal points in logarithms to which correspond *negative* numbers have been replaced by the letter *n*.

The demonstrations that have been made now enable us to see at once from an inspection of the results in this table that

$$z_4 = 0.238$$
$$p_{z_4} = 1.6$$
$$[vv] = .007$$

Therefore substituting in equations (73) and (74) we have

$$\mu_{z_4} = .047 \qquad r_{z_4} = .032$$

If the two values of $[vv]$ obtained in the solution had differed at all we should have taken the mean of the two.

113. If the Elimination Equations (95) are divided by $[aa]$, $[bb, 1]$, $[cc, 2]$, $[dd, 3]$, respectively, they become

$$\left.\begin{aligned} z_1 + A_b z_2 + A_c z_3 + A_d z_4 + A_m &= 0 \\ z_2 + B_c z_3 + B_d z_4 + B_m &= 0 \\ z_3 + C_d z_4 + C_m &= 0 \\ z_4 + D_m &= 0 \end{aligned}\right\} \quad (99)$$

And the solution for the unknowns can be effected most conveniently by arranging the computations in the manner illustrated on page 109.

Scheme A. Solution of the Normal Equations.

	a	b	c	d	m	s	δ
a	5	−3	−4	−3	−1.3	−6.3	0
b		8	1	1	− .1	6.9	0
A_b	−.6	1.8	2.4	1.8	.78	3.78	
c			5	0	.9	2.9	0
A_c	−.8		3.2	2.4	1.04	5.04	
d				8	1.1	7.1	0
A_d	−.6			1.8	.78	3.78	
m					.64	1.24	0
A_m	−.26				.34	1.64	
b		6.2	−1.4	− .8	− .88	3.12	0
		0.7924	0_n1461	9_n9031	9_n9445	0.4924	
c			1.8	−2.4	− .14	−2.14	0
			.316	.181	.199	− .705	
B_c	9_n3537		9.4998	9.2568	9.2982	9_n8479	
d				6.2	.32	3.32	0
				.103	.114	− .403	
B_d	9_n1107			9.0138	9.0552	9_n6049	
m					.30	− .40	0
					.125	− .443	
B_m	9_n1521				9.0966	9_n6463	
c			1.484	−2.581	− .339	−1.435	1
			0.1715	0_n4118	9_n5302	0_n1569	
d				6.097	.206	3.373	1
				4.489	.589	2.496	
C_d	0_n2403			0.6521	9.7705	0.3972	
m					.175	.043	1
					.077	.328	
C_m	9_n3587				8.8889	9.5156	
d				1.608	− .383	1.227	2
				0.2063	9_n5832	0.0888	
m					.098	− .285	0
					.091	− .292	
D_m	$9_n3769 = \log - z_4$				8.9601	9_n4657	
m	$0.238 = z_4$			$[vv] = .007$	.007	.007	0

SCHEME B.— SOLUTION OF THE ELIMINATION EQUATIONS.

$-D_m$	$-C_m$	$-B_m$	$-A_m$
	$-C_d\,z_4$	$-B_d\,z_4$	$-A_d\,z_4$
		$-B_c\,z_3$	$-A_c\,z_3$
			$-A_b\,z_2$
z_4	z_3	z_2	z_1
$\log z_4$	$\log z_3$	$\log z_2$	
$\log C_d$	$\log B_c$	$\log A_b$	
$\log B_d$	$\log A_c$		
$\log A_d$			
$\log C_d\,z_4$	$\log B_c\,z_3$	$\log A_b\,z_2$	
$\log B_d\,z_4$	$\log A_c\,z_3$		
$\log A_d\,z_4$			

114. Filling out this table for the example just solved, we have

.238	.228	.142	.260
	.414	.031	.143
		.145	.514
			.191
.238	.642	.318	1.108
9.3769	9.8075	9.5024	
0_n2403	9_n3537	9_n7782	
9_n1107	9_n9031		
9_n7782			
9_n6172	9_n1612	9_n2806	
8_n4876	9_n7106		
9_n1551			

115. The Weights of the Unknowns. In order to determine the precision measures of z_1, z_2, and z_3, it would next be necessary to compute the weights of the latter quantities. The demonstration of the processes by which these weights may be found will not be taken up here, as the best method to adopt varies a good deal with the character of the example, but a statement of the results in the general form of solution will be given.

By treating the Elimination Equations in a way similar to that used in deriving equation (E) of paragraph 76 from equations (B) of the same paragraph, we may show that

$$\begin{array}{llllll} z_1 + A_m & + B_m \alpha_1 & + C_m \alpha_2 & + D_m \alpha_3 & = 0 & \\ z_2 & + B_m & + C_m \beta_2 & + D_m \beta_3 & = 0 & (100) \\ z_3 & & + C_m & + D_m \gamma_3 & = 0 & \end{array}$$

where the α's, β's, γ's, are determined from the equations

$$\begin{array}{lll} A_d + B_d \alpha_1 + C_d \alpha_2 + \alpha_3 = 0 & B_d + C_d \beta_2 + \beta_3 = 0 & \\ A_c + B_c \alpha_1 + \alpha_2 = 0 & B_c + \beta_2 = 0 & (101) \\ A_b + \alpha_1 = 0 & C_d + \gamma_3 = 0 & \end{array}$$

Then by an application of the principles of Rule 1, paragraph (77), it may be shown that

$$\begin{aligned} \frac{1}{p_{z_1}} &= \frac{1}{[aa]} + \frac{\alpha_1^2}{[bb, 1]} + \frac{\alpha_2^2}{[cc, 2]} + \frac{\alpha_3^2}{[dd, 3]} \\ \frac{1}{p_{z_2}} &= \frac{1}{[bb, 1]} + \frac{\beta_2^2}{[cc, 2]} + \frac{\beta_3^2}{[dd, 3]} \\ \frac{1}{p_{z_3}} &= \frac{1}{[cc, 2]} + \frac{\gamma_3^2}{[dd, 3]} \qquad (102) \\ \frac{1}{p_{z_4}} &= \frac{1}{[dd, 3]} \end{aligned}$$

These equations can of course be extended to cover any number of unknown quantities, and tabular schemes for the computations of the α's, β's, γ's, . . . and of the weights p_{z_1}, p_{z_2}, p_{z_3}, . . . can readily be arranged.

For a general demonstration of these results, and also for a discussion of special methods of solution, consult—

Johnson, "The Theory of Errors and Method of Least Squares," chap. ix.

Wright, "Treatise on the Adjustment of Observations," chap. iv.

Chauvenet, "Spherical and Practical Astronomy," pp. 530–549.

THE METHOD OF CORRELATIVES.

116. The method of adjusting "Conditioned Observations" explained in paragraph 33 is perfectly general, but where there are many conditions to be satisfied the solution is apt to be very laborious. For the case that occurs most frequently in practice, in which the observations are direct and equal in number to the number of unknown quantities, the process of solution devised by *Gauss* and called the "Method of Correlatives" is the most convenient. This method is derived as follows:—

Let q observations, M_1, M_2, . . . M_q, of the respective weights p_1, p_2, . . . p_q, be made directly upon the values of q unknown quantities, and let the most probable values of the unknowns be

$$z_1 = M_1 + v_1, \quad z_2 = M_2 + v_2, \quad \ldots z_q = M_q + v_q.$$

Where v_1, v_2, . . . v_q, are the most probable corrections to apply to the observed values as well as in this case the residuals of the observations.

If the n' condition equations are not linear they may be reduced to that form by the method of paragraph 44, so that we may assume for our

CONDITION EQUATIONS

$$\left.\begin{array}{l} a_1v_1 + a_2v_2 + \ldots a_qv_q + m_1 = 0 \\ b_1v_1 + b_2v_2 + \ldots b_qv_q + m_2 = 0 \\ \ldots \quad \ldots \quad \ldots \\ \ldots \quad \ldots \quad \ldots \\ l_1v_1 + l_2v_2 + \ldots l_qv_q + m_{n'} = 0 \end{array}\right\} \quad \text{(A)}$$

In which the quantities $m_1, m_2, \ldots m_{n'}$, would all be zero if the observations were exact. It is to be observed that the coefficients $a, b, \ldots l$ are not arranged in the same order in these equations as they are in the observation equations of paragraph 107.

The values of $v_1, v_2, \ldots v_q$, must be determined so as to satisfy the above equations and also by the principle of Least Squares, so as to make

$$p_1v_1^2 + p_2v_2^2 + \ldots p_qv_q^2 = \text{a minimum.}$$

Corresponding with a minimum value of this last we have

$$p_1v_1dv_1 + p_2v_2dv_2 + \ldots p_qv_qdv_q = 0 \qquad \text{(B)}$$

for all possible simultaneous values of $dv_1, dv_2, \ldots dv_q$; that is, for all values which satisfy the equations,

$$\left.\begin{array}{l} a_1dv_1 + a_2dv_2 + \ldots a_qdv_q = 0 \\ b_1dv_1 + b_2dv_2 + \ldots b_qdv_q = 0 \\ \ldots \quad \ldots \quad \ldots \\ \ldots \quad \ldots \quad \ldots \\ l_1dv_1 + l_2dv_2 + \ldots l_qdv_q = 0 \end{array}\right\} \quad \text{(C)}$$

obtained by differentiating equations (A).

Therefore, denoting the first member of (B) by R and the first members of (C) by $S_1, S_2, \ldots S_{n'}$, it will be necessary that

$$R - k_1S_1 - k_2S_2 - \ldots k_{n'}S_{n'} = 0 \qquad \text{(D)}$$

where $k_1, k_2, \ldots k_{n'}$, are undetermined coefficients.

This last equation will be satisfied if the coefficient of each differential in it is made equal to zero, that is, if

$$
\begin{aligned}
p_1 v_1 &= k_1 a_1 + k_2 b_1 + \dots k_{n'} l_1 \\
p_2 v_2 &= k_1 a_2 + k_2 b_2 + \dots k_{n'} l_2 \\
&\dots \qquad \dots \qquad \dots \\
&\dots \qquad \dots \qquad \dots \\
p_q v_q &= k_1 a_q + k_2 b_q + \dots k_{n'} l_q
\end{aligned}
\qquad (103)
$$

All that remains therefore is to find values of $v_1, v_2, \dots v_q$ and $k_1, k_2, \dots k_{n'}$, which will satisfy simultaneously equations (A) and (103), and that this may be done is easily seen from the fact that we have the same number of equations as unknowns.

Substituting the values of $v_1, v_2, \dots v_q$ from equations (103) in equations (A) we have the following:

$$
\begin{aligned}
k_1 \Sigma \frac{aa}{p} + k_2 \Sigma \frac{ab}{p} + \dots k_{n'} \Sigma \frac{al}{p} + m_1 &= 0 \\
k_1 \Sigma \frac{ab}{p} + k_2 \Sigma \frac{bb}{p} + \dots k_{n'} \Sigma \frac{bl}{p} + m_2 &= 0 \\
\dots \qquad \dots \qquad \dots & \\
\dots \qquad \dots \qquad \dots & \\
k_1 \Sigma \frac{al}{p} + k_2 \Sigma \frac{bl}{p} + \dots k_{n'} \Sigma \frac{ll}{p} + m_{n'} &= 0
\end{aligned}
\qquad (104)
$$

The solution of these equations gives at once the values of $k_1, k_2, \dots k_{n'}$, which are called the "Correlatives" of the Condition Equations. The values of $v_1, v_2, \dots v_q$ are then found by substituting the values of the k's in equation (103).

117. As equations (104) are of the same general form as a set of Normal Equations, *Gauss's* Method of Substitution can be advantageously employed in the solution.

When there is but a single equation of condition the second members of equations (103) reduce to their first terms, and equations (104) reduce to the single equation

$$k_1 \Sigma \frac{aa}{p} + m_1 = 0 \qquad (105)$$

and the values of $v_1, v_2, \ldots v_q$ in (103) become

$$v_1 = -\frac{\frac{a_1}{p_1}}{\Sigma \frac{aa}{p}} m_1, \quad v_2 = -\frac{\frac{a_2}{p_2}}{\Sigma \frac{aa}{p}} m_1, \ldots$$

$$v_q = -\frac{\frac{a_q}{p_q}}{\Sigma \frac{aa}{p}} m_1 \qquad (106)$$

It is from these results that the rules in paragraph 35 are derived.

118. *Example.* Suppose we have given the observations

$$\begin{aligned} M_1 &= 2.02, &&\text{weight } 3 \\ M_2 &= 4.13, &&\text{“} \quad 2 \\ M_3 &= 2.52, &&\text{“} \quad 5 \\ M_4 &= 2.67, &&\text{“} \quad 7 \\ M_5 &= 2.84, &&\text{“} \quad 4 \end{aligned} \qquad (a)$$

and let the most probable values of the unknowns be represented by

$$z_1 = M_1 + v_1, \quad z_2 = M_2 + v_2, \ldots z_5 = M_5 + v_5 \quad (b)$$

Also suppose that the unknowns are subject to the conditions

$$\begin{aligned} z_1 + z_2 + z_3 + z_4 + z_5 &= 14.0 \\ z_2 \qquad - z_4 \qquad &= 1.5 \end{aligned} \qquad (c)$$

Then expressing these conditions in terms of the corrections by means of (a) and (b), we have the

CONDITION EQUATIONS

$$\begin{aligned} v_1 + v_2 + v_3 + v_4 + v_5 + .18 &= 0 \\ v_2 \qquad - v_4 \qquad - .04 &= 0 \end{aligned} \qquad \text{(d)}$$

Referring to paragraph 116, we see that in this example

$$n' = 2, \qquad m_1 = 0.18, \qquad m_2 = -0.04$$

For the purpose of computing the coefficients in equations (104) we next arrange the following table.

p	a	b	$\frac{aa}{p}$	$\frac{ab}{p}$	$\frac{bb}{p}$
3	1	0	$\frac{1}{3}$	0	0
2	1	1	$\frac{1}{2}$	$\frac{1}{2}$	$\frac{1}{2}$
5	1	0	$\frac{1}{5}$	0	0
7	1	-1	$\frac{1}{7}$	$-\frac{1}{7}$	$\frac{1}{7}$
4	1	0	$\frac{1}{4}$	0	0
			$\frac{599}{420}$	$\frac{5}{14}$	$\frac{9}{14}$

Substituting these results in equations (104), we have

$$\left.\begin{aligned}\frac{599}{420}k_1 + \frac{5}{14}k_2 + .18 &= 0\\ \frac{5}{14}k_1 + \frac{9}{14}k_2 - .04 &= 0\end{aligned}\right\}\qquad(e)$$

Solving,

$$\left.\begin{aligned}k_1 &= -.1647\\ k_2 &= .1537\end{aligned}\right\}\qquad(f)$$

Then from equations (103) we get at once

$$\left.\begin{aligned}v_1 &= -.0549, & v_4 &= -.0455,\\ v_2 &= -.0055, & v_5 &= -.0412.\\ v_3 &= -.0329,\end{aligned}\right\}\qquad(g)$$

And by substituting these results in (b) we can obtain the values of z_1, z_2, z_3, z_4, z_5.

The above is the solution of Example 70, page 127.

EXAMPLES.

1. An urn contains five black balls, three red balls and two white balls. If three balls are drawn from the urn what different combinations may result, and what is the probability of each?

2. In a single throw with a pair of dice what is the probability that neither ace nor doublets will appear? $\frac{5}{9}$

3. Four cards are drawn from a pack. What is the probability of getting four aces? Of getting one of each suit?

4. From a lottery of thirty tickets, marked 1, 2, . . . 30, four tickets are drawn. What is the probability that numbers 1 and 15 will be among them? $\frac{2}{145}$

5. Find the odds against the appearance of 7 or 11 in a single throw with a pair of dice. 7 : 2

6. I toss up n coins. What is my chance of getting just one head?

7. In a single throw what are the relative chances of throwing 9 with two dice and with three dice? 24 : 25

8. From 2 n counters marked with consecutive numbers two are drawn. What are the odds against having an even sum? $n : n - 1$

9. In two trials with a single die what is the probability of throwing (a) an ace the first time only? (b) at least one ace?

10. Find the probability of throwing doublets one or more times in three trials with a pair of dice. $\frac{91}{216}$

11. Find the probability of throwing exactly three aces in five trials with a single die. $\frac{125}{3888}$

12. A certain stake is to be won by the first person who throws 5 with a die of twelve faces. What is the chance of the sixth person?

13. A and B play chess. A wins on the average two games out of three. What is A's chance of winning just four games out of the first six? $\frac{80}{243}$

14. A and B shoot alternately at a mark. A hits once in n times and B once in $n-1$ times. Find their chances of first hit, and the odds in favor of B if A misses on his first shot. Even. $n : n-2$

15. In how many trials will it be a wager of 4 to 3 that double five will be thrown with a pair of dice? 30

16. Find the probability of throwing one and only one ace in two trials with a single die. $\frac{5}{18}$

17. If I have three tickets in a lottery of four prizes and eight blanks, what is my chance of drawing a prize? $\frac{41}{55}$

18. Find the probability of throwing at least four aces in six trials with a single die. $\frac{203}{23328}$

19. On an average seven ships out of eight return to port. Find the chance that out of five ships expected at least three will return. $\frac{16121}{16384}$

20. In a lottery containing a large number of tickets, where the prizes are to the blanks as 1 : 9, find the chance of drawing at least two prizes in five trials. $\frac{8146}{100000}$

21. In a purse are ten coins, all nickels except one which is a five-dollar gold piece; in another are ten coins, all nickels. Nine coins are taken from the first purse and placed in the second, and then nine coins are taken from the latter and placed in the former. If you now had your choice which purse would you take?

22. A and B engage in a game in which A's skill is to B's as 2 : 3. What is A's chance of winning at least two games out of five?

23. If A's skill at a certain game is double that of B, what are the odds against A's winning four games before B wins two? 131 : 112

24. A party of twenty-five take seats at a round table. What are the odds against any two specified persons sitting next to each other?

25. A has three shares in a lottery in which there are three prizes and six blanks. B has one share in another where there is but one prize and two blanks. What are their relative chances of getting a prize? A : B = 16 : 7

26. Expand through the terms involving h^3 and k^3, the expression

$$\frac{1}{x+h} + (y+k)^3$$

When x is 1 and y is $\frac{1}{2}$, does $\frac{1}{x} + y^3$ increase or diminish when x and y begin to increase at the same rate?

27. Given $f(x, y) = x^2(a + y)^3$, expand the expression $(x + h)^2 (a + y + k)^3$.

28. Find the value of $\log_{10} a + \cos b$, when $a = 1001$ and $b = 0.1°$. Give the result first to five places and then to seven places of significant figures, in each case without the aid of tables. 4.000433

29. Transform to the new origin, $(2, -3, 1)$, the equation, $x^2 + y^2 + z^2 - 4x + 6y - 2z - 11 = 0$, the

axes remaining parallel to the original ones. The equations of transformation are $x = 2 + x'$, $y = -3 + y'$, $z = 1 + z'$. $x^2 + y^2 + z^2 = 25$

30. Find the minimum value of $x^2 + xy + y^2 - ax - by$. $\frac{(ab - a^2 - b^2)}{3}$

31. Find the values of x, y and z that render a maximum or a minimum the function $x^2 + y^2 + z^2 + x - 2z - xy$.

32. Find the values of x and y that render a maximum or a minimum the expression $\sin x + \sin y + \cos (x + y)$.

33. Find the co-ordinates of a point the sum of the squares of whose distances from three given points, (x_1, y_1), (x_2, y_2), (x_3, y_3), is a minimum. $x = \frac{(x_1 + x_2 + x_3)}{3}$

34. Given the volume, a^3, of a rectangular parallelopiped, find its shape when its surface is a minimum.

35. Find the volume of the greatest rectangular parallelopiped that can be inscribed in the ellipsoid

$$\frac{x^2}{a^2} + \frac{y^2}{b^2} + \frac{z^2}{c^2} = 1 \qquad \frac{8abc}{3\sqrt{3}}$$

36. In the Physical Laboratory apparatus for illustrating the Estimation of Tenths, the reading of the micrometer head for a certain setting is computed to be 2.3038, which may be taken as exact. Setting the apparatus by the eye the following readings are obtained :—

2.314	2.324	2.310	2.519	2.326
2.320	2.302	2.313	2.305	

What are the accidental and real errors and the residuals? Are there any constant errors or mistakes? If there are constant errors are they of the first, second, or third class? How do you tell?

37. Are the following observations such as to call for an application of the Method of Least Squares in their adjustment?

$$x + y - z + u = 5 \qquad 3x - y - z - u = 1$$
$$x - 2y + 2z + u = -1 \qquad x - y + 4z + 6u = 9$$

38. In the case of direct observations, what other quantities besides the arithmetical mean might reasonably be assumed to give plausible values of the unknowns? Why is the arithmetical mean preferred to these?

39. Find the most probable value of a quantity M from the observations

216.27	216.16	216.04	216.19	216.44	216.58
.29	.43	215.99	.39	.51	
.33	.09	216.23	.14	215.94	

Also test the result by finding the sum of the residuals.

$M = 216.251$

40. In the determination of a certain wave length, Rowland made the following observations. Find the most probable wave length.

4.524	4.515	4.513	4.507	4.501	4.485	4.517	4.493	4.505
.500	.508	.511	.497	.502	.519	.504	.492	

4.5055 ± 0.0017

41. Ten measurements of the density of a body gave results as follows. Find the most probable density.

9.662	9.664	9.677	9.663	9.645
.673	.659	.662	.680	.654

9.6639 ± 0.0022

42. In a triangulation of the U. S. Coast Survey an angle was measured twenty-four times with results

116° 43′ 44″.45	48″.90	47″.40	47″.85	51″.75
50 .55	49 .20	47 .75	50 .60	49 .00
50 .95	48 .85	51 .05	48 .45	52 .35

51″.30	49″.05	46″.75
51 .05	50 .55	49 .25
51 .70	49 .25	53 .40

Find the best value for the angle. 116° 43′ 49″.64 ± 0″.28

43. The following observations were made with a sextant in order to determine the latitude of a place:—

43° 4′ 46″	7″	59″	52″	47″	36″
24	28	39	52	15	40

What is the most probable latitude? 43° 4′ 27″ ± 5″

44. Determine the quantity M from the observations

$M =$	81.55	.41	.68	.25	.27	.77	.13	.86	.10	.03
Wts.	25	25	16	9	9	9	4	1	1	1

45. An angle M is measured with the following results:

M	p	M	p	M	p	M	p	M	p
45″.00	5	42″.50	5	27″.50	3	36″.25	2	45″.00	2
31 .25	4	37 .50	3	43 .33	3	42 .50	3	40 .83	3
45 .00	3	38 .33	3	40 .63	4	39 .17	3		

Find the most probable value of the angle and test the result by computing Σpv. 39″.78 ± 0″.94

46. In one hundred measurements of angles made in the primary triangulation in Massachusetts by the Coast Survey there were found between

6″.0	and	5″.0	1 error.	1″.0	and	0″.0	26 errors.
5 .0	"	4 .0	2 errors.	0 .0	"	− 1 .0	26 "
4 .0	"	3 .0	2 "	− 1 .0	"	− 2 .0	17 "
3 .0	"	2 .0	3 "	− 2 .0	"	− 3 .0	8 "
2 .0	"	1 .0	13 "	− 3 .0	"	− 4 .0	2 "

Plot these results as in Figure 1, page 8.

47. In sixty-six determinations of the velocity of light made at Washington the percentage of errors of different magnitudes was found to be as follows: —

Over	− 7.0	.8%	Over	7.0	0.0%
Equal to	− 7.0	.8	Equal to	7.0	1.6
	− 6.0	1.6		6.0	3.5
	− 5.0	2.0		5.0	4.2
	− 4.0	3.9		4.0	5.8
	− 3.0	7.8		3.0	8.6
	− 2.0	12.8		2.0	12.4
	− 1.0	17.1		1.0	17.1

Plot the results and draw a smooth curve.

48. Given the observations

$$x - y + 2z = 3 \qquad 4x + y + 4z = 21$$
$$3x + 2y - 5z = 5 \qquad -x + 3y + 3z = 14$$

find the most probable values of x, y, and z.

$x = 2.470 \pm .038$

49. O being the level of the sea and P_1, P_2, P_3, three points whose altitudes are to be determined, the following observations are made: —

P_1 above $O = 10$ ft. $\qquad P_2$ above $P_3 = 9$ ft.
P_2 " $P_1 = 7$ $\qquad P_1$ " $P_3 = 2$
P_2 " $O = 18$

Find the most probable altitudes. $P_3 = 8.50 \pm .29$

50. The altitudes of A above O, B above A, and B above O are found by measurements to be respectively 12.3, 14.1, and 27.0 feet. What is the most probable value of each of these differences in level? $A = 12.50 \pm .17$

51. Measurements of the ordinates of points on a straight line corresponding to abscissas 4, 6, 8, 9, are made with results 5, 8, 10, 12. What is the most probable equation of the line in the form $y = mx + b$? $b = -0.29$

52. Find the altitudes in Example 49 if the observations have weights 5, 3, 6, 2, 4, respectively.

53. Solve the example in paragraph 31, giving the observations the weights 25, 25, 4, 4, 4, 4, 4, 4, 1, respectively.

Elevation of $P_5 = 320.25$

54. Find the most probable values of z_1, z_2, and z_3 from the observations

$z_1 = 552.10$	wt. 16	$z_1 - z_2 = .75$	wt. 1
$-z_2 + z_3 = .15$	" 9	$z_1 + z_2 - z_3 = 552.05$	" 1
$z_3 = 551.23$	" 4	$z_1 - z_3 = .70$	" 1
$z_2 = 551.30$	" 4		

$z_2 = 551.2345$

55. In the triangulation of Lake Superior there were measured at station O the angles

$FOP =$	62°	59′	40″.33	wt. 5
$FOR =$	64	11	34 .92	" 7
$FOB =$	100	20	29 .12	" 4
$POB =$	37	20	49 .55	" 7
$ROB =$	36	8	55 .86	" 4

Required the adjusted values of the angles.

$FOP = 40''.28 \pm 0''.34$

56. In the U. S. Lake Survey the following angles were measured at station North Base :—

(1) Crebassa — Middle	55°	57′	58″.68	wt.	3
(2) Middle — Quaquaming	48	49	13 .64	"	19
(3) Crebassa — Quaquaming	104	47	12 .66	"	17
(4) Quaquaming — South Base	54	38	15 .53	"	13
(5) Middle — South Base	103	27	28 .99	"	6

Find the adjusted values of the angles.

$(1) = 58''.965;\ r = 0''.28$

57. Adjust the following observations of differences in level: —

Altitude of A	401.3	wt. 16	C above B	72.5	wt. 9
A above B	220.8	" 16	A " B	222.0	" 1
A " C	150.2	" 4	Altitude of B	180.7	" 1

58. In "Conditioned Observations" can the number of observations required be less than the number of unknown quantities? Why must the number of conditions be less than the number of unknowns?

59. From the following measurements of the angles formed at the centre of a disk by four radial lines, find the most probable values of the angles.

$$A = 104^\circ\ 25'\ 13'' \qquad C = 86^\circ\ 33'\ 20''$$
$$B = 98\ \ 13\ \ 47 \qquad D = 70\ \ 48\ \ 23$$
$$A = 104^\circ\ 25'\ 2''.25$$

Also solve giving the observations the weights 5, 2, 1, 4, respectively.

60. Four observations on the angle A of a triangle gave a mean of 36° 25′ 47″, two observations on B gave a mean of 90° 36′ 28″, and three on C gave 52° 57′ 57″. Adjust the triangle. $A = 36^\circ\ 25'\ 44''.2;\ r = 7''.7$

61. Five angles at a station are measured, and also their sum. The observed sum differs from the sum of the five observed parts by the amount d. What are the adjusted values of the angles?

62. The three angles of a spherical triangle are measured with results

$$A = 46^\circ\ 17'\ 38''.32 \qquad B = 73^\circ\ 35'\ 16''.15$$
$$C = 60^\circ\ 7'\ 5''.16.$$

Adjust the triangle, knowing that the spherical excess is 2″.475. $A = 39''.3;\ \mu = 1''.6$

63. At the station Pine Mountain the following angles were observed between surrounding stations: —

Jocelyne — Deepwater	65°	11′	52″.500	wt.	3
Deepwater — Deakyne	66	24	15 .553	"	3
Deakyne — Burden	87	2	24 .703	"	3
Burden — Jocelyne	141	21	21 .757	"	1

Find the most probable values of the angles.

64. Solve Examples 55 and 56 by the method of "Conditioned Observations."

65. A is a station whose altitude is known to be 5240.1 feet. B and C are floats on a lake, and D is a signal point. From the following observations determine the most probable altitudes of B, C and D.

C below	A	720.1	wt. 3		B below	A	719.7	wt. 3
D "	A	200.3	" 5		B "	D	520.9	" 2
C "	D	520.4	" 2					

66. Given the following observations, subject to the condition $z_1 + z_2 = z_3$, find the most probable values of z_1, z_2, and z_3.

$$2z_1 - z_2 + z_3 = 3.0 \qquad 2z_2 - z_3 = 1.0$$
$$2z_1 - 3z_2 = -4.5 \qquad z_1 + 2z_2 = 5.1$$
$$z_1 + z_3 = 3.8$$

67. The chemical composition of a specimen was found by several observers to be as follows: —

$Pb = .52$	Other substances $= .09$	Au and $Ag = .39$
$Ag = .27$	Pb and $Ag = .78$	Impurities $= .10$
$Au = .11$	Pb and impurities $= .62$	$Au = .12$

From these observations find the most probable composition of the specimen.

68. From the following observations what are the best values of the unknowns, supposing that y and z must be equal?

$x + y = 5.2$	wt. 4		$y + z = 4.2$	wt. 1
$x = 3.0$	" 9		$z = 2.0$	" 4
$x - z = 1.1$	" 1			

69. In determining the difference in longitude between various cities the results obtained were

(1)	Cambridge — Washington	23^m	$41^s.041$	wt. 30
(2)	Cambridge — Cleveland	42	14 .875	" 7
(3)	Cambridge — Columbus	47	27 .713	" 8
(4)	Washington — Columbus	23	46 .816	" 7
(5)	Cleveland — Columbus	5	12 .929	" 5

Adjust these observations.

70. The capacity of a condenser is known to be 14.0 m. f. It is divided into five sections, a, b, c, d, e, and it is known that the difference between b and d is 1.5 m. f. Find the most probable capacities of the sections from the observations

$a = 2.02$	wt. 3	$d = 2.67$	wt. 7
$b = 4.13$	" 2	$e = 2.84$	" 4
$c = 2.52$	" 5		

$a = 1.9651$

71. If the unknowns in the following observations are subject to the condition $x + 2y + 3z = 36$, what are their adjusted values?

$x = 4.3$ wt. 1, $y = 5.7$ wt. 4, $z = 7.3$ wt. 9

$x = 3.77$

72. A cannon is discharged horizontally from the top of a bluff. Observations on the time, and distance of fall of the ball gave the results

$t =$	0.5	1.0	1.5	2.0	seconds
$s =$	1.2	4.0	9.1	15.0	metres

What curve, passing through the point of departure of the ball, will represent the above observations?

73. An Argand burner shows the following efficiencies with varying rates of gas consumption:

g =	2.0	2.3	2.8	3.3	4.0	4.5	5.0	feet
E =	2.1	2.4	2.5	3.0	3.2	3.8	4.1	

Find the equation of the straight line which best represents the relation between g and E. The measurements on g are without appreciable error.

74. Observations are made upon the expansion of Amyl alcohol with change in temperature as follows:—

V =	1.04	1.12	1.19	1.24	1.27	cu. cm.
t =	13.9	43.0	67.8	89.0	99.2	C. degrees

If $V = 1 + B\,t + C\,t^2$ expresses the law connecting the volume and temperature, find the most probable values of B and C.

75. In a Hooke's joint where the angle between the axes is 45°, x being the angular rotation of the driver, and y that of the follower, from the following measurements find the equation of a curve that will represent the relation between x and $y - x$.

x	$y - x$	x	$y - x$	x	$y - x$
0°	0°.0	80°	− 5°.8	140°	8°.8
20°	− 5°.7	90°	− 2°.0	160°	5°.3
40°	− 9°.9	100°	2°.3	180°	0°.0
60°	− 10°.4	120°	8°.0		

$$y - x = -0.85 - 9.82 \sin 2x + 0.92 \cos 2x$$

76. A series of observations extending over a period of thirty years was made by Quetelet to determine the daily variation in temperature at Brussels. The mean results of

the measurements are given below. From them derive an equation to express the temperature at any time of the year.

Jan.	4°.66	May	9°.83	Sept.	8°.16
Feb.	5°.42	June	10°.09	Oct.	6°.55
Mar.	6°.77	July	9°.71	Nov.	5°.10
Apr.	8°.59	Aug.	9°.14	Dec.	4°.41

$$y = 7.369 + 0.9854 \sin 30x - 2.7084 \cos 30x + 0.0100 \sin 60x - 0.1950 \cos 60x - 0.0133 \sin 90x + 0.1783 \cos 90x$$

In this answer the values of x begin at the 15th of January, and represent the time in months.

77. The law connecting the time of vibration of a pendulum with its length is assumed to be of the form, $T = m L^n$. From the following observations find the most probable values of m and n.

$T =$	12.9	11.6	10.4	9.7	5.3	4.6
$L =$	164.4	132.9	107.6	93.5	28.4	20.6

L is in centimetres, T in tenths seconds. $n = 0.5000$

$m = 1.0044$

78. Determine the equation of a curve which will represent the following observations:

x	y	x	y	x	y
0.0	0.00	1.5	1.09	3.0	8.65
0.5	0.04	2.0	2.56	3.5	13.72
1.0	0.31	2.5	4.99	4.0	20.47

79. Determine the equation of a curve which will represent the relation between x and y in the observations,

x	y	x	y	x	y	x	y
0.0	4.51	0.3	4.09	0.6	3.03	1.2	0.92
0.1	4.44	0.4	3.76	0.8	2.24	1.5	0.38
0.2	4.31	0.5	3.42	1.0	1.49	2.0	0.05

80. At a station P the angles between a straight line passing through P parallel to the axis of X and the directions from P of four points P_1, P_2, P_3, P_4, are measured. Having given the coördinates, (a, b), of the four points, find the coördinates of P.

Point.	*Coördinates.*		*Angle.*
	a	b	
P_1	4.21	3.24	39° 18′
P_2	1.21	2.10	147° 54′
P_3	− 0.51	0.22	205° 24′
P_4	2.50	− 1.10	277° 15′

If the coördinates of the point P are (x, y), and the angle is denoted by A, we have

$$\tan A = \frac{y - b}{x - a}$$

81. If $a \sin bx = M$, and values of M are observed for known values of a and b, determine the most probable value of x. If x' is an approximate value of x found by trial, and $m = a \sin bx' - M$, we shall have

$$x = x' - \frac{\Sigma\, a\, b\, m \cos bx'}{\Sigma\,(a\, b \cos bx')^2}$$

82. If in one series of observations the value of h is twice what it is in another, what is the relative probability of the occurrence of an error of given magnitude a in the two series? Show what the curves of error will be in the two cases. What error has the same probability for its occurrence in each series? What is the relative probability of the occurrence of an error not greater than a in the first case and not greater than $2a$ in the second case?

83. From 64 observations the latitude of a station was found to be 49° 10′ 9″.110 ± 0″.051. What was the probable error of a single observation? 0″.41

84. If twenty measurements of an angle give a result with an *A.D.* of $0''.38$, and it is required to find the angle so that the *A.D.* shall be only $0''.25$, how many more observations must be made? 27

85. From the following determinations of the area of a field find the most probable area and its probable error.

$$5674 \pm 12, \quad 5680 \pm 4, \quad 5685 \pm 3, \quad 5682 \pm 1, \quad 5678 \pm 2$$

$$A = 5681.41 \pm 0.84$$

86. From the following measurements by *Fizeau* and others, find the most probable value for the velocity of light together with its probable error. Measurements are in kilometers.

298000 ± 1000	299990 ± 200	299930 ± 100
298500 ± 1000	300100 ± 1000	

$$V = 299917 \pm 88$$

87. Two different instruments give for the value of an angle,

$$34° \; 55' \; 33''.0 \pm 4''.1, \qquad 34° \; 55' \; 36''.0 \pm 6''.3$$

What is the best value to take for the angle?

$$34° \; 55' \; 33''.9 \pm 3''.4$$

88. Determinations of the difference in longitude between Washington and Key West made on seven different days gave the results

19^m	$1^s.42 \pm 0^s.044$	19^m	$1^s.60 \pm 0^s.046$
	$1\;.37 \pm 0\;.037$		$1\;.55 \pm 0\;.045$
	$1\;.38 \pm 0\;.036$		$1\;.57 \pm 0\;.047$
	$1\;.45 \pm 0\;.036$		

What is the best value and its probable error?

$$1^s.460 \pm 0^s.016$$

89. In the triangulation of Lake Ontario two different instruments gave for an angle, 74° 25′ 5″.429 ± 0″.29 from sixteen readings, and 74° 25′ 4″.611 ± 0″.22 from twenty-four readings. Find the most probable value of the angle and its probable error.

90. In each of Examples 39–45 find the mean and probable errors and average deviation of each observation and of the most probable value, using formulas from (50) to (63) according as they apply.

91. In Example 42 divide the observations in their order into six groups of four observations each and compute the mean of each group. Then determine the probable error of the first of these means: (1) considered as a single measure of four times the weight of those in Example 42; (2) directly as one of six observations of equal weight; (3) as a determination from its four constituents. 0″.67; 0″.72; 1″.00

92. The following twenty-nine measurements on the density of the earth, made by Cavendish, give as a mean result 5.48. What is the probable error of an observation? Solve by the usual method and also by taking the residual that occupies the middle position. 0.14

5.50	5.55	5.57	5.34	5.42	5.30
.61	.36	.53	.79	.47	.75
.88	.29	.62	.10	.63	.68
.07	.58	.29	.27	.34	.85
.26	.65	.44	.39	.46	

93. What is the probable error of the mean of two observations which differ by the amount a?

94. A base-line is measured five times with a steel tape reading to hundredths of a foot, and five times with a chain reading to tenths of a foot, with results

By tape,	741.17	741.09	741.22	741.12	741.10
By chain,	741.2	741.4	741.0	741.3	741.1

Find the probable errors and weights for a single observation in each case, and also the adjusted length of the line and its probable error. 741.146 ± 0.015

95. Twenty-one determinations of a chronometer correction gave results

− 8.78	− 8.78	− 8.68	− 8.80	− 8.96	− 8.83	− 8.79
.76	.51	.63	.75	.64	.70	.90
.85	.64	.58	.78	.65	.64	.93

Find the probable error of the mean by using both formulas (53) and (57), and also determine the probable error of a single observation by taking the middle residual.

0.017; 0.018; 0.09

96. In the following observations show that $M_0 = 49.64$, $\mu = 1.95$, $r = 1.31$, $\mu_0 = 0.40$, $r_0 = 0.27$, $\mu_3 = 0.87$, $r_3 = 0.59$.

$M =$ 48.81	48.76	49.53	51.56	50.38	49.84
$p =$ 5	4	5	3	2	5

97. Observations on the time of ending of a transit of Mercury are made by different observers with a variety of instruments and under more or less favorable circumstances. If the weights assigned by the computer are as indicated, find the best value for the time and its probable error.

$5^h\ 38^m\ 23^s$	wt. 1	$38^m\ 26^s$	wt. 3	$38^m\ 19^s$	wt. 3
37 55	" 0	38 21	" 2	38 21	" 2
38 10	" 1	38 18	" 2	38 15	" 2

$t_0 = 5^h\ 38^m\ 19^s.9$

98. An angle is measured five times with a theodolite, and seven times with a transit, giving results

Theodolite, 31″.7, 39″.8, 40″.7, 28″.6, 32″.3
Transit, 32 .8, 36 .7, 38 .2, 29 .3, 41 .6 35″.3, 36″.2

If the relative values of readings by the two instruments are as 3 to 2, what is the most probable value of the angle? What is the mean error of the result?

99. Given $M_1 = 65.58 \pm .59$, $M_2 = 35.15 \pm .93$, $M_3 = 49.64 \pm .27$, find the probable errors of $4\,M_1 - 3\,M_2 + 2\,M_3$ and of $\frac{M_1}{2} + \frac{M_2}{3} - \frac{M_3}{4}$. 3.69; 0.43

100. The three angles of a triangle are measured, and the probable error of each observation is r. What is the probable error of the triangle error? $r\sqrt{3}$

101. The zenith distance of a star on the meridian is observed to be $z = 21^\circ\ 17'\ 20''.3 \pm 2''.3$. The declination of the star is given as $d = 19^\circ\ 30'\ 14''.8 \pm 0''.8$. What is the latitude of the place and its probable error?

$$L = z + d = 40^\circ\ 47'\ 35''.1 \pm 2''4.$$

102. The zenith distance z of a star at upper culmination is observed n times, and its zenith distance z' at lower culmination n' times. If the latitude is given by $L = 90^\circ - \frac{1}{2}(z + z')$, and the probable error of an observation is r, what is the probable error of the latitude?

103. The horizontal force necessary to start a 100-pound weight sliding along a table is observed to be 15.5 ± 0.2 pounds. Find the probable error of the coefficient of friction.

104. If a line is measured by the continued application of a unit of measure, and r is the probable error of the placing and reading of this measure, what is the probable error of the length l? $r\sqrt{l}$

105. If the average deviations of z_1, z_2, z_3, are a, b, c, respectively, what is the average deviation of $z_1^2 + z_2^2 + z_3^2$?

106. If the radius of a circle is measured with result 1000.0 ± 2.0, how should the circumference and area be expressed?

107. Two sides, a and b, and the included angle C of a triangle are measured with results $a = 252.52 \pm .06$

feet, $b = 300.01 \pm .06$ feet, $C = 42° \; 13' \; 00'' \pm 30''$. What is the area and its probable error? 25452 ± 9

108. Measurements of adjacent sides of a rectangle gave $a \pm r_1$, and $b \pm r_2$. What is the probable error of the area, and for what kind of a rectangle will this probable error be the least?

109. If the measured sides of a rectangle have the same *a.d.*, what is the *a.d.* of the diagonal determined from them? Same

110. If the sides of a rectangle are measured in the manner indicated in Example 104 and found to be a and b, what is the probable error of the area?

111. The correction to be applied to a chronometer is found to be $+ 12^m \, 13^s.2 \pm 0^s.3$. Ten days later the correction is again determined and found to be $12^m \, 21^s.4 \pm 0^s.3$. What is the mean daily rate and its probable error?

$0^s.820 \pm 0^s.042$

112. Measurements of the compression of the earth's meridian have resulted in

$$\frac{1}{294} \pm .000046$$

What is the probable error of the denominator 294? 3.98

113. The current flowing in a circuit is due to two sources whose electromotive forces are determined to be $e_1 = 200 \pm 2$, $e_2 = 400 \pm 3$. The resistance of the circuit is 30 ± 1. Find the current and its probable error.

20 ± 0.68

114. The side b and angles B and C of a triangle are measured with results $b = 106 \pm .06$ metres, $B = 29° \; 39' \pm 1'$, $C = 120° \; 7' \pm 2'$. What is the most probable value of the angle A and of the side c?

$A = 30° \; 14' \pm 2'.2$; $c = 185.35 \pm 0.15$

115. The distance between two divisions on a graduated scale is measured by a micrometer. Show that the average

deviation of the mean of two results is the same as the average deviation of a single reading.

116. If the weights of the determinations of three angles A, B, C, are 3, 3, 1, respectively, what is the weight of the sum of the three angles? 0.6

117. If the weight of x is p, what is the weight of $\log_a x$?

118. If $x = \frac{y}{c}$ and the weight of y is p, what is the weight of x? c^2p

119. In Example 107, how closely must the parts be measured in order to obtain the area within 0.5 per cent?

120. From observations on l and t the value of g is to be computed by the pendulum formula

$$t = \pi\sqrt{\frac{l}{g}}$$

What changes in g will be produced by changes in l and t of δ_1 and δ_2 units, respectively, and what are the allowable errors in l and t if g is to be determined within 1 per cent?

121. The moment of inertia of a cylindrical bar is to be obtained from measurements on its mass m, its length h, and its diameter d. The error in the determination of m is negligible, the precision of the determination of d is four times that of h. If the measurements give $m = 48$, $h = 8.000$, $d = 1.200 \pm 0.10$, and

$$I = m\left(\frac{h^2}{12} + \frac{d^2}{16}\right)$$

what is the probable error of I, and what should be the ratio of d to h to determine I most accurately?

$d : h = 256 : 9$

122. If observations give for a certain quantity x the value 303, with a mean error of 2, what is the mean error of the expression $3x + \log_{10} 2x$?

123. The probable error of the determination of the angle A is $20''$. What is the maximum probable error of $\sin A + \cos A$?

124. If the probable error of an observation on an angle is $10''$, is there any difference between the probable error of the function $\sin A + \cos A + \sin C$ and of the function $\sin A + \cos B + \sin C$, supposing A and B are of the same magnitude?

125. Given the observations,

$$z_1 - 2z_2 + z_3 - 3 = 0 \qquad 3z_1 + z_2 + 2z_3 - 17 = 0$$
$$2z_1 + 3z_2 - 4z_3 - 2 = 0 \qquad -z_1 + 4z_2 + 3z_3 - 10 = 0$$

Find the most probable values of z_1, z_2, z_3, and also their weights and precision measures.

$$z_1 = 3.541\,; \; p_{z_1} = 29\,; \; r_{z_1} = .024$$

126. Find the weights and precision measures of the unknowns in Examples 48 to 57.

127. Determine the probable errors of the constants in Examples 72 to 79, inclusive.

128. The length of a pendulum which beats seconds is given by

$$l = l' + \left(\frac{5}{2} q - s\right) l' \sin^2 L$$

where l' is the length at the equator, q the ratio $\frac{1}{289}$ of the centrifugal force at the equator to the weight, and s the compression of the meridian regarded as unknown. Putting

$$l' = 991^{m.m.} + x, \qquad \left(\frac{5}{2} q - s\right) l' = y,$$

observations in different latitudes gave in millimetres the

following equations, from which we are to determine l and s together with their probable errors: —

$$\begin{array}{llll} x + 0.969y = 5.13 & & x + 0.152y = 0.77 \\ x + 0.749y = 3.97 & & x + 0.327y = 1.70 \\ x + 0.426y = 2.24 & & x + 0.685y = 3.62 \\ x + 0.095y = 0.56 & & x + 0.793y = 4.23 \\ x \qquad\quad\;\; = 0.19 & & \end{array}$$

$$l' = 991.069 \pm .026; \quad s = \frac{1}{294} \pm 0.00046$$

129. Find the weights and precision measures of the unknowns in Examples 64 to 70, and also in Examples 59 to 63, and in 71.

130. In Example 95 the probable error of a single observation is 0.08 seconds. Find the number of errors which should fall between 0.00 seconds and 0.10 seconds, between 0.10 seconds and 0.20 seconds, and also the number that should be over 0.20 seconds. Compare the results with the number actually found.

131. In 470 determinations of the right ascensions of Sirius and Altair made by Bradley, the probable error of a single observation was 0″.2637. The number of errors falling between specified limits was as shown below. Compare this result with the distribution of errors called for by the theory.

Limits.	*Errors.*	*Limits.*	*Errors.*
0″.0 to 0″.1	94	0″.6 to 0″.7	26
0 .1 to 0 .2	88	0 .7 to 0 .8	14
0 .2 to 0 .3	78	0 .8 to 0 .9	10
0 .3 to 0 .4	58	0 .9 to 1 .0	7
0 .4 to 0 .5	51	Over 1 .0	8
0 .5 to 0 .6	36		

132. What is the probability that the error of a single observation will be as large as twice the probable error? As large as five times the probable error?

133. On the average how many observations must be made before an error as large as three times the mean error will occur?

134. In Example 46, assuming that all errors between any two limits fall half way between those limits, compute the average deviation and mean error of an observation and compare their ratio with the theoretical value given in the table in paragraph 55.

135. A line is measured 500 times and the probable error of each observation is 0.6 c.m. How many errors should occur between 0.4 c.m. and 0.8 c.m.?

136. Show how the value of π could be determined experimentally from observations such as those in Example 131.

137. In a system of observations all equally good, r being the probable error of a single observation, if two observations are taken at random, what quantity is their difference as likely as not to exceed, and what is the probability that the difference will be less than r? $r\sqrt{2}$; 0.367

138. In the following measurements of an angle, ought any of the observations to be rejected?

12′ 51″.75	47″.85	47″.40	48″.90	44″.45
48 .45	51 .05	48 .85	50 .95	
50 .60	47 .75	49 .20	50 .55	

139. Determine whether any of the observations in Example 44 should be rejected.

140. A quantity M is measured with the results given below. Ought all the observations to be retained?

$$M = 236,\ 251,\ 249,\ 252,\ 248,\ 254,\ 246,\ 257,\ 243,\ 274$$

141. A certain angle has been laid out with such accuracy that its true value may be taken as exactly 90°. Twenty-five observations are made upon it with a transit that it is

desired to test, and the result obtained is $89° 59' 57'' \pm 0''.8$. What are the odds in favor of a constant error in the instrument between $-1''$ and $-5''$? Between $0''$ and $-6''$?

908 : 92; 86 : 1

142. Repeated measurements of a standard metre bar with a decimetre scale gave a result 10.032 ± 0.010. What are the odds in favor of a constant error in the scale between $\frac{0.010}{\sqrt{10}}$ and $\frac{0.054}{\sqrt{10}}$? 43 : 7

143. Two determinations of the length of a line gave 683.4 ± 0.3 and 684.9 ± 0.3, respectively. Show that the best value for the length is 684.15 ± 0.51, and that the probable systematic error of each determination is 0.65.

144. Two men A and B observe an angle repeatedly with the same instrument with results

A.		*B.*	
47° 23′ 40″	23′ 35″	47° 23′ 30″	24′ 00″
23 45	23 40	23 40	23 20
23 30		23 50	

Is there any relative personal error, and what is the best final value? $47° 23' 38''.2 \pm 1''.6$

145. Three independent determinations of the capacity of a condenser made with three different instruments gave results 42.22 ± 0.21, $43.40 \pm .15$ and 44.20 ± 0.18. What is the most probable value of the capacity?

For extended treatment of the subject illustrated in Examples 143 to 145 see *Johnson*, "The Theory of Errors and Method of Least Squares," chap. vii.

146. In an estimation of tenths what is the probable error of an observation? What is the average deviation? 0.025

147. In obtaining the angle of deflection of the needle of a tangent galvanometer by the usual method what is the probable error of the result?

148. If all the errors of a series of observations must fall between 0 and a, and the frequency of any error is proportional to its magnitude, what is the Curve of Error? What are the values of r, $a.d.$, and μ? $r = \frac{a}{\sqrt{2}}$

149. In Example 148 what is the probability that the error of a single observation will be as large as $0.5\ a$.

150. If all values of x between 0 and a are possible, and their probabilities are proportional to their squares, find the mean value of x and the probability that x will be as large as $0.5\ a$. Also draw the Curve of Error.

151. What is the greatest probable error of a logarithm found by interpolation in a seven place table? .000000015

152. Given the following set of Normal Equations, together with $[mm] = 1.3409$, find the most probable values of the unknowns and their weights and probable errors. There were sixteen observations.

$$3.1217\,z_1 + .5756\,z_2 - .1565\,z_3 - .0067\,z_4 - 1.5710 = 0$$
$$.5756\,z_1 + 2.9375\,z_2 + .1103\,z_3 - .0015\,z_4 + .9275 = 0$$
$$-.1565\,z_1 + .1103\,z_2 + 4.1273\,z_3 + .2051\,z_4 + .0652 = 0$$
$$-.0067\,z_1 - .0015\,z_2 + .2051\,z_3 + 4.1328\,z_4 + .0178 = 0$$

$$z_1 = 0.583 \pm 0.018$$
$$z_4 = -0.004 \pm 0.015$$
$$p_{z_4} = 4.12$$

153. From ten observation equations, for which was found $[mm] = 2.6322$, there resulted the normal equations

$$5.2485\,z_1 - 1.7472\,z_2 - 2.1954\,z_3 + 0.5399 = 0$$
$$-1.7472\,z_1 + 1.8859\,z_2 + 0.8041\,z_3 - 1.4493 = 0$$
$$-2.1954\,z_1 + 0.8041\,z_2 + 4.0440\,z_3 - 1.8681 = 0$$

Find the most probable values of z_1, z_2 and z_3 together with their probable errors. $z_1 = 0.42 \pm 0.11$

154. Find the most probable values of the unknowns in the normal equations

$$\begin{aligned}
459\, z_1 - 308\, z_2 - 389\, z_3 + 244\, z_4 - 507 &= 0 \\
-\ 308\, z_1 + 464\, z_2 + 408\, z_3 - 269\, z_4 + 695 &= 0 \\
-\ 389\, z_1 + 408\, z_2 + 676\, z_3 - 331\, z_4 + 653 &= 0 \\
244\, z_1 - 269\, z_2 - 331\, z_3 + 469\, z_4 - 283 &= 0 \\
[mm] &= 1129
\end{aligned}$$

$$z_4 = -\ 0.488\,;\ p_{z_4} = 281$$

155. If thirteen observation equations give rise to the result $[mm] = 100.34$ and to the normal equations

$$\begin{aligned}
17.50\, z_1 - 6.50\, z_2 - 6.50\, z_3 - 2.14 &= 0 \\
-\ 6.50\, z_1 + 17.50\, z_2 - 6.50\, z_3 - 13.96 &= 0 \\
-\ 6.50\, z_1 - 6.50\, z_2 + 20.50\, z_3 + 5.40 &= 0
\end{aligned}$$

show that the most probable values of the unknowns are

$$z_1 = 0.67 \pm 0.60, \quad z_2 = 1.17 \pm 0.60, \quad z_3 = 0.32 \pm 0.55$$

APPENDIX.

ELEMENTS OF THE THEORY OF PROBABILITY.

200. Definition. If an event can happen in a ways, and fail in b ways, and all these ways are equally likely to occur, the probability of the happening of the event is $\frac{a}{a+b}$, and the probability of its failure is $\frac{b}{a+b}$.

Since the event must either happen or fail, the sum of the above probabilities must represent a certainty. But

$$\frac{a}{a+b} + \frac{b}{a+b} = 1$$

That is, the probability of a certainty is expressed by unity. Also, if the probability, P, of the happening of an event is known, the probability of its failure is given at once by $1 - P$.

201. *Example A.* A single throw is made with a pair of dice. What is the probability that the sum of the spots turned up will be 5 ?

Number of ways of throwing the dice is $6 \times 6 = 36$

Number of ways of throwing five is 4

$\therefore$ Probability of throwing five is $\frac{4}{36} = \frac{1}{9}$

Example B. A coin is tossed up six times. Find the chance that three heads and three tails will be the result.

Number of ways of throwing the coin is $2^6 = 64$

Number of ways of throwing three heads is $\dfrac{6 \times 5 \times 4}{1 \times 2 \times 3} = 20$

Probability of throwing three heads is $\dfrac{20}{64} = \dfrac{5}{16}$

202. Compound Events. A certain event can happen in a ways and fail in b ways: a second independent event can happen in a' ways, and fail in b' ways, all of these ways being equally likely to occur.

To find the probability of the simultaneous occurrence of the two events.

The total number of ways in which the events can take place together is $(a + b)(a' + b')$

(1) Both events can happen in $a\,a'$ ways.
(2) Both events can fail in $b\,b'$ ways.
(3) First event can happen and second fail in $a\,b'$ ways.
(4) First event can fail and second happen in $a'\,b$ ways.

$\therefore$ The probability of (1) is $\dfrac{a\,a'}{(a + b)(a' + b')}$

The probability of (2) is $\dfrac{b\,b'}{(a + b)(a' + b')}$

The probability of (3) is $\dfrac{a\,b'}{(a + b)(a' + b')}$

The probability of (4) is $\dfrac{a'\,b}{(a + b)(a' + b')}$

But the probability of the happening of the first event is $\dfrac{a}{a + b}$, and of the second event is $\dfrac{a'}{a' + b'}$, etc. Hence it will at once be seen that the probability of the simultaneous occurrence of two independent events is equal to the product of the probabilities of the occurrence of the component events.

Or, in general, if $P_1, P_2, \ldots P_n$ are the probabilities of the occurrence of any number, n, of independent events, the probability of the simultaneous occurrence of all the events is

$$P_1 \times P_2 \times \ldots P_n \qquad \text{(A)}$$

By independent events is meant those such that the manner of occurrence of one has no influence upon the manner of occurrence of the others.

203. *Example C.* The chance that A can solve a certain problem is $\frac{2}{3}$, and the chance that B can solve it is $\frac{5}{12}$. Find,

(a) The probability that both will solve it.
(b) The probability that the problem will be solved.

For (a). This is a question as to the probability of the concurrence of two independent events. Therefore by an application of (A), the probability that both will solve the problem is

$$\frac{2}{3} \times \frac{5}{12} = \frac{5}{18}$$

For (b). The problem will be solved unless both fail.

The probability that both will fail is $\frac{1}{3} \times \frac{7}{12} = \frac{7}{36}$

$\therefore$ The probability of getting a solution is $1 - \frac{7}{36} = \frac{29}{36}$

Example D. A pack of cards is cut, and those taken off then replaced. In how many trials will it be an even wager that an ace will be cut?

Let n be the number of trials. Then n is to be found from

$$1 - \left(\frac{48}{52}\right)^n = \frac{1}{2}$$

where the first member of the equation represents the probability that we shall not fail n times in succession.

Solving for n,

$$n = \frac{\log 2}{\log 52 - \log 48}$$
$$= 8.7$$

In nine trials then there is a little more than an even chance of cutting an ace.

204. Dependent Events. If we have a number of events whose modes of occurrence are dependent one upon another, the probability of their concurrence will be found by the same method as in paragraph 202; a' now denoting the number of ways in which after the first event has happened the second will follow, and b' the number of ways in which after the first has happened the second will not follow, etc. Accordingly, the general formula (A) of paragraph 202 applies to dependent events as well as to independent ones.

Also, if an event can take place in a variety of ways, the total probability of its occurrence will be the sum of the probabilities of its occurrence in each of the different ways.

205. *Example E.* Suppose two purses contain respectively five dimes and a copper, and six dimes. A coin is taken at random from the first purse and placed in the second, and then a coin is transferred from the second to the first. What is the probability that the copper will remain in the first purse?

The probability that the copper will be taken from the first purse and placed in the second, and then returned to the first purse is

$$\frac{1}{6} \times \frac{1}{7} = \frac{1}{42}$$

and the probability that the copper will not be taken from the first purse at all is

$$\frac{5}{6}$$

Therefore the probability that the copper will finally remain in the first purse is

$$\frac{1}{42} + \frac{5}{6} = \frac{36}{42} = \frac{6}{7}$$

FUNCTIONS OF SEVERAL VARIABLES.

206. For the application of Taylor's Theorem to the expansion of a function of several independent variables, see *Osborne's* "Differential and Integral Calculus," page 145. And for the conditions that lead to maxima and minima values of such functions, see page 155 of the same work.

BIBLIOGRAPHY.

The following brief list of treatises, dealing with the Method of Least Squares, is appended for the benefit of those whose professional work requires such constant application of the process as to render desirable a more detailed knowledge of various special methods of solution. In connection with some of the titles attention is called to the subjects the treatment of which is particularly full.

Johnson, "The Theory of Errors and Method of Least Squares."
Probability of Errors. Systematic Errors. The Method of Substitution.

Wright, "Treatise on the Adjustment of Observations."
Special Methods of Solution. Applications to Geodetic and Engineering Problems.

Merriman, "Text-Book on the Method of Least Squares."

Chauvenet, "Treatise on the Method of Least Squares."
Development of the Theory. Applications to Astronomical Observations.

Bobek, "Lehrbuch der Ausgleichsrechnung nach der Methode der Kleinsten Quadrate."
General Synopsis of the Method, illustrated by Numerous Examples.

Koll, "Die Methode der Kleinsten Quadrate."
Applications to Geodesy.

Hansen, "Von der Methode der Kleinsten Quadrate."
Applications to Geodesy.

Helmert, "Die Ausgleichungsrechnung nach der Methode der Kleinsten Quadrate."

Liagre, "Calcul des Probabilités."

Holman, "Discussion of the Precision of Measurements."
Problems in Physics and Electrical Engineering.

Weinstein, "Handbuch der Physikalischen Maassbestimmungen."
Applications to Physical Problems.

Oppolzer, "Lehrbuch zur Bahnbestimmung der Kometen und Planeten."

Jordan, "Handbuch der Vermessungskunde."

For a complete list of works on the Method of Least Squares published up to 1876, see

Merriman, "A List of Writings relating to the Method of Least Squares, with Historical and Critical Notes."
Published in the Transactions of the Connecticut Academy, vol. iv, 1877.

Notice of works published since 1876 may be found in periodicals devoted to the progress of Mathematical Science. Such as

"Jahrbuch über die Fortschritte der Mathematik."

"Bulletin des Sciences Mathématiques."

TABLE I.

Values of the Integral $\frac{2}{\sqrt{\pi}}\int_0^t e^{-t^2}dt$ for Argument $\frac{t}{0.4769}$ or $\frac{a}{r}$

$\frac{a}{r}$	0	1	2	3	4	5	6	7	8	9	Diff.
0.0	0.0000	0.0054	0.0108	0.0161	0.0215	0.0269	0.0323	0.0377	0.0430	0.0484	54
0.1	0538	0591	0645	0699	0752	0806	0859	0913	0966	1020	54
0.2	1073	1126	1180	1233	1286	1339	1392	1445	1498	1551	53
0.3	1603	1656	1709	1761	1814	1866	1918	1971	2023	2075	52
0.4	2127	2179	2230	2282	2334	2385	2436	2488	2539	2590	51
0.5	0.2641	0.2691	0.2742	0.2793	0.2843	0.2893	0.2944	0.2994	0.3043	0.3093	50
0.6	3143	3192	3242	3291	3340	3389	3438	3487	3535	3583	49
0.7	3632	3680	3728	3775	3823	3870	3918	3965	4012	4059	46
0.8	4105	4152	4198	4244	4290	4336	4381	4427	4472	4517	45
0.9	4562	4606	4651	4695	4739	4783	4827	4860	4914	4957	43
1.0	0.5000	0.5043	0.5085	0.5128	0.5170	0.5212	0.5254	0.5295	0.5337	0.5378	41
1.1	5419	5460	5500	5540	5581	5620	5660	5700	5739	5778	39
1.2	5817	5856	5894	5932	5970	6008	6046	6083	6120	6157	37
1.3	6194	6231	6267	6303	6339	6375	6410	6445	6480	6515	35
1.4	6550	6584	6618	6652	6686	6719	6753	6786	6818	6851	32
1.5	0.6883	0.6915	0.6947	0.6979	0.7011	0.7042	0.7073	0.7104	0.7134	0.7165	30
1.6	7195	7225	7255	7284	7313	7342	7371	7400	7428	7457	28
1.7	7485	7512	7540	7567	7594	7621	7648	7675	7701	7727	26
1 8	7753	7778	7804	7829	7854	7879	7904	7928	7952	7976	24
1.9	8000	8023	8047	8070	8093	8116	8138	8161	8183	8205	22
2.0	0.8227	0.8248	0.8270	0.8291	0.8312	0.8332	0.8353	0.8373	0.8394	0.8414	19
2 1	8433	8453	8473	8492	8511	8530	8549	8567	8585	8604	18
2.2	8622	8639	8657	8674	8692	8709	8726	8742	8759	8775	17
2 3	8792	8808	8824	8840	8855	8870	8886	8901	8916	8930	15
2.4	8945	8960	8974	8988	9002	9016	9029	9043	9056	9069	13
2.5	0.9082	0.9095	0.9108	0.9121	0.9133	0.9146	0.9158	0.9170	0.9182	0.9193	12
2.6	9205	9217	9228	9239	9250	9261	9272	9283	9293	9304	10
2.7	9314	9324	9334	9344	9354	9364	9373	9383	9392	9401	9
2.8	9410	9419	9428	9437	9446	9454	9463	9471	9479	9487	8
2.9	9495	9503	9511	9519	9526	9534	9541	9548	9556	9563	7
3 0	0.9570	0.9577	0.9583	0.9590	0.9597	0.9603	0.9610	0.9616	0.9622	0.9629	6
3.1	9635	9641	9647	9652	9658	9664	9669	9675	9680	9686	5
3.2	9691	9696	9701	9706	9711	9716	9721	9726	9731	9735	5
3.3	9740	9744	9749	9753	9757	9761	9766	9770	9774	9778	4
3.4	9782	9786	9789	9793	9797	9800	9804	9807	9811	9814	4
3.	0.9570	0.9635	0.9691	0.9740	0.9782	0.9818	0.9848	0.9874	0.9896	0.9915	
4.	9930	9943	9954	9963	9970	9976	9981	9985	9988	9990	
5.	9993	9994	9996	9997	9997	9998	9998	9999	9999	9999	
∞	1.0000										
$\frac{a}{r}$	0	1	2	3	4	5	6	7	8	9	Diff.

TABLE II.—Common Logarithms.

n	0	1	2	3	4	5	6	7	8	9	Diff.
10	0000	0043	0086	0128	0170	0212	0253	0294	0334	0374	42
11	0414	0453	0492	0531	0569	0607	0645	0682	0719	0755	38
12	0792	0828	0864	0899	0934	0969	1004	1038	1072	1106	35
13	1139	1173	1206	1239	1271	1303	1335	1367	1399	1430	32
14	1461	1492	1523	1553	1584	1614	1644	1673	1703	1732	30
15	1761	1790	1818	1847	1875	1903	1931	1959	1987	2014	28
16	2041	2068	2095	2122	2148	2175	2201	2227	2253	2279	27
17	2304	2330	2355	2380	2405	2430	2455	2480	2504	2529	25
18	2553	2577	2601	2625	2648	2672	2695	2718	2742	2765	24
19	2788	2810	2833	2856	2878	2900	2923	2945	2967	2989	22
20	3010	3032	3054	3075	3096	3118	3139	3160	3181	3201	21
21	3222	3243	3263	3284	3304	3324	3345	3365	3385	3404	20
22	3424	3444	3464	3483	3502	3522	3541	3560	3579	3598	19
23	3617	3636	3655	3674	3692	3711	3729	3747	3766	3784	18
24	3802	3820	3838	3856	3874	3892	3909	3927	3945	3962	18
25	3979	3997	4014	4031	4048	4065	4082	4099	4116	4133	17
26	4150	4166	4183	4200	4216	4232	4249	4265	4281	4298	17
27	4314	4330	4346	4362	4378	4393	4409	4425	4440	4456	16
28	4472	4487	4502	4518	4533	4548	4564	4579	4594	4609	15
29	4624	4639	4654	4669	4683	4698	4713	4728	4742	4757	15
30	4771	4786	4800	4814	4829	4843	4857	4871	4886	4900	14
31	4914	4928	4942	4955	4969	4983	4997	5011	5024	5038	14
32	5051	5065	5079	5092	5105	5119	5132	5145	5159	5172	13
33	5185	5198	5211	5224	5237	5250	5263	5276	5289	5302	13
34	5315	5328	5340	5353	5366	5378	5391	5403	5416	5428	13
35	5441	5453	5465	5478	5490	5502	5514	5527	5539	5551	12
36	5563	5575	5587	5599	5611	5623	5635	5647	5658	5670	12
37	5682	5694	5705	5717	5729	5740	5752	5763	5775	5786	12
38	5798	5809	5821	5832	5843	5855	5866	5877	5888	5899	11
39	5911	5922	5933	5944	5955	5966	5977	5988	5999	6010	11
40	6021	6031	6042	6053	6064	6075	6085	6096	6107	6117	11
41	6128	6138	6149	6160	6170	6180	6191	6201	6212	6222	11
42	6232	6243	6253	6263	6274	6284	6294	6304	6314	6325	10
43	6335	6345	6355	6365	6375	6385	6395	6405	6415	6425	10
44	6435	6444	6454	6464	6474	6484	6493	6503	6513	6522	10
45	6532	6542	6551	6561	6571	6580	6590	6599	6609	6618	10
46	6628	6637	6646	6656	6665	6675	6684	6693	6702	6712	9
47	6721	6730	6739	6749	6758	6767	6776	6785	6794	6803	9
48	6812	6821	6830	6839	6848	6857	6866	6875	6884	6893	9
49	6902	6911	6920	6928	6937	6946	6955	6964	6972	6981	9
50	6990	6998	7007	7016	7024	7033	7042	7050	7059	7067	9
51	7076	7084	7093	7101	7110	7118	7126	7135	7143	7152	8
52	7160	7168	7177	7185	7193	7202	7210	7218	7226	7235	8
53	7243	7251	7259	7267	7275	7284	7292	7300	7308	7316	8
54	7324	7332	7340	7348	7356	7364	7372	7380	7388	7396	8
n	0	1	2	3	4	5	6	7	8	9	Diff.

TABLE II.— Common Logarithms.

n	0	1	2	3	4	5	6	7	8	9	Diff.
55	7404	7412	7419	7427	7435	7443	7451	7459	7466	7474	8
56	7482	7490	7497	7505	7513	7520	7528	7536	7543	7551	
57	7559	7566	7574	7582	7589	7597	7604	7612	7619	7627	
58	7634	7642	7649	7657	7664	7672	7679	7686	7694	7701	
59	7709	7716	7723	7731	7738	7745	7752	7760	7767	7774	
60	7782	7789	7796	7803	7810	7818	7825	7832	7839	7846	7
61	7853	7860	7868	7875	7882	7889	7896	7903	7910	7917	
62	7924	7931	7938	7945	7952	7959	7966	7973	7980	7987	
63	7993	8000	8007	8014	8021	8028	8035	8041	8048	8055	
64	8062	8069	8075	8082	8089	8096	8102	8109	8116	8122	
65	8129	8136	8142	8149	8156	8162	8169	8176	8182	8189	7
66	8195	8202	8209	8215	8222	8228	8235	8241	8248	8254	
67	8261	8267	8274	8280	8287	8293	8299	8306	8312	8319	
68	8325	8331	8338	8344	8351	8357	8363	8370	8376	8382	
69	8388	8395	8401	8407	8414	8420	8426	8432	8439	8445	
70	8451	8457	8463	8470	8476	8482	8488	8494	8500	8506	6
71	8513	8519	8525	8531	8537	8543	8549	8555	8561	8567	
72	8573	8579	8585	8591	8597	8603	8609	8615	8621	8627	
73	8633	8639	8645	8651	8657	8663	8669	8675	8681	8686	
74	8692	8698	8704	8710	8716	8722	8727	8733	8739	8745	
75	8751	8756	8762	8768	8774	8779	8785	8791	8797	8802	6
76	8808	8814	8820	8825	8831	8837	8842	8848	8854	8859	
77	8865	8871	8876	8882	8887	8893	8899	8904	8910	8915	
78	8921	8927	8932	8938	8943	8949	8954	8960	8965	8971	
79	8976	8982	8987	8993	8998	9004	9009	9015	9020	9025	
80	9031	9036	9042	9047	9053	9058	9063	9069	9074	9079	5
81	9085	9090	9096	9101	9106	9112	9117	9122	9128	9133	
82	9138	9143	9149	9154	9159	9165	9170	9175	9180	9186	
83	9191	9196	9201	9206	9212	9217	9222	9227	9232	9238	
84	9243	9248	9253	9258	9263	9269	9274	9279	9284	9289	
85	9294	9299	9304	9309	9315	9320	9325	9330	9335	9340	5
86	9345	9350	9355	9360	9365	9370	9375	9380	9385	9390	
87	9395	9400	9405	9410	9415	9420	9425	9430	9435	9440	
88	9445	9450	9455	9460	9465	9469	9474	9479	9484	9489	
89	9494	9499	9504	9509	9513	9518	9523	9528	9533	9538	
90	9542	9547	9552	9557	9562	9566	9571	9576	9581	9586	5
91	9590	9595	9600	9605	9609	9614	9619	9624	9628	9633	
92	9638	9643	9647	9652	9657	9661	9666	9671	9675	9680	
93	9685	9689	9694	9699	9703	9708	9713	9717	9722	9727	
94	9731	9736	9741	9745	9750	9754	9759	9763	9768	9773	
95	9777	9782	9786	9791	9795	9800	9805	9809	9814	9818	4
96	9823	9827	9832	9836	9841	9845	9850	9854	9859	9863	
97	9868	9872	9877	9881	9886	9890	9894	9899	9903	9908	
98	9912	9917	9921	9926	9930	9934	9939	9943	9948	9952	
99	9956	9961	9965	9969	9974	9978	9983	9987	9991	9996	
n	0	1	2	3	4	5	6	7	8	9	Diff.

TABLE III.—Squares of Numbers.

n	0	1	2	3	4	5	6	7	8	9	Diff.
1.0	1.000	1.020	1.040	1.061	1.082	1.103	1.124	1.145	1.166	1.188	22
1.1	1.210	1.232	1.254	1.277	1.300	1.323	1.346	1.369	1.392	1.416	24
1.2	1.440	1.464	1.488	1.513	1.538	1.563	1.588	1.613	1.638	1.664	26
1.3	1.690	1.716	1.742	1.769	1.796	1.823	1.850	1.877	1.904	1.932	28
1.4	1.960	1.988	2.016	2.045	2.074	2.103	2.132	2.161	2.190	2.220	30
1.5	2.250	2.280	2.310	2.341	2.372	2.403	2.434	2.465	2.496	2.528	32
1.6	2.560	2.592	2.624	2.657	2.690	2.723	2.756	2.789	2.822	2.856	34
1.7	2.890	2.924	2.958	2.993	3.028	3.063	3.098	3.133	3.168	3.204	36
1.8	3.240	3.276	3.312	3.349	3.386	3.423	3.460	3.497	3.534	3.572	38
1.9	3.610	3.648	3.686	3.725	3.764	3.803	3.842	3.881	3.920	3.960	40
2.0	4.000	4.040	4.080	4.121	4.162	4.203	4.244	4.285	4.326	4.368	42
2.1	4.410	4.452	4.494	4.537	4.580	4.623	4.666	4.709	4.752	4.796	44
2.2	4.840	4.884	4.928	4.973	5.018	5.063	5.108	5.153	5.198	5.244	46
2.3	5.290	5.336	5.382	5.429	5.476	5.523	5.570	5.617	5.664	5.712	48
2.4	5.760	5.808	5.856	5.905	5.954	6.003	6.052	6.101	6.150	6.200	50
2.5	6.250	6.300	6.350	6.401	6.452	6.503	6.554	6.605	6.656	6.708	52
2.6	6.760	6.812	6.864	6.917	6.970	7.023	7.076	7.129	7.182	7.236	54
2.7	7.290	7.344	7.398	7.453	7.508	7.563	7.618	7.673	7.728	7.784	56
2.8	7.840	7.896	7.952	8.009	8.066	8.123	8.180	8.237	8.294	8.352	58
2.9	8.410	8.468	8.526	8.585	8.644	8.703	8.762	8.821	8.880	8.940	60
3.0	9.000	9.060	9.120	9.181	9.242	9.303	9.364	9.425	9.486	9.548	62
3.1	9.610	9.672	9.734	9.797	9.860	9.923	9.986	10.05	10.11	10.18	6
3.2	10.24	10.30	10.37	10.43	10.50	10.56	10.63	10.69	10.76	10.82	7
3.3	10.89	10.96	11.02	11.09	11.16	11.22	11.29	11.36	11.42	11.49	7
3.4	11.56	11.63	11.70	11.76	11.83	11.90	11.97	12.04	12.11	12.18	7
3.5	12.25	12.32	12.39	12.46	12.53	12.60	12.67	12.74	12.82	12.89	7
3.6	12.96	13.03	13.10	13.18	13.25	13.32	13.40	13.47	13.54	13.62	7
3.7	13.69	13.76	13.84	13.91	13.99	14.06	14.14	14.21	14.29	14.36	8
3.8	14.44	14.52	14.59	14.67	14.75	14.82	14.90	14.98	15.05	15.13	8
3.9	15.21	15.29	15.37	15.44	15.52	15.60	15.68	15.76	15.84	15.92	8
4.0	16.00	16.08	16.16	16.24	16.32	16.40	16.48	16.56	16.65	16.73	8
4.1	16.81	16.89	16.97	17.06	17.14	17.22	17.31	17.39	17.47	17.56	8
4.2	17.64	17.72	17.81	17.89	17.98	18.06	18.15	18.23	18.32	18.40	9
4.3	18.49	18.58	18.66	18.75	18.84	18.92	19.01	19.10	19.18	19.27	9
4.4	19.36	19.45	19.54	19.62	19.71	19.80	19.89	19.98	20.07	20.16	9
4.5	20.25	20.34	20.43	20.52	20.61	20.70	20.79	20.88	20.98	21.07	9
4.6	21.16	21.25	21.34	21.44	21.53	21.62	21.72	21.81	21.90	22.00	9
4.7	22.09	22.18	22.28	22.37	22.47	22.56	22.66	22.75	22.85	22.94	10
4.8	23.04	23.14	23.23	23.33	23.43	23.52	23.62	23.72	23.81	23.91	10
4.9	24.01	24.11	24.21	24.30	24.40	24.50	24.60	24.70	24.80	24.90	10
5.0	25.00	25.10	25.20	25.30	25.40	25.50	25.60	25.70	25.81	25.91	10
5.1	26.01	26.11	26.21	26.32	26.42	26.52	26.63	26.73	26.83	26.94	10
5.2	27.04	27.14	27.25	27.35	27.46	27.56	27.67	27.77	27.88	27.98	11
5.3	28.09	28.20	28.30	28.41	28.52	28.62	28.73	28.84	28.94	29.05	11
5.4	29.16	29.27	29.38	29.48	29.59	29.70	29.81	29.92	30.03	30.14	11
n	0	1	2	3	4	5	6	7	8	9	Diff.

TABLE III.— Squares of Numbers.

n	0	1	2	3	4	5	6	7	8	9	Diff.
5.5	30.25	30.36	30.47	30.58	30.69	30.80	30.91	31.02	31.14	31.25	11
5.6	31.36	31.47	31.58	31.70	31.81	31.92	32.04	32.15	32.26	32.38	11
5.7	32.49	32.60	32.72	32.83	32.95	33.06	33.18	33.29	33.41	33.52	12
5.8	33.64	33.76	33.87	33.99	34.11	34.22	34.34	34.46	34.57	34.69	12
5.9	34.81	34.93	35.05	35.16	35.28	35.40	35.52	35.64	35.76	35.88	12
6.0	36.00	36.12	36.24	36.36	36.48	36.60	36.72	36.84	36.97	37.09	12
6.1	37.21	37.33	37.45	37.58	37.70	37.82	37.95	38.07	38.19	38.32	12
6.2	38.44	38.56	38.69	38.81	38.94	39.06	39.19	39.31	39.44	39.56	13
6.3	39.69	39.82	39.94	40.07	40.20	40.32	40.45	40.58	40.70	40.83	13
6.4	40.96	41.09	41.22	41.34	41.47	41.60	41.73	41.86	41.99	42.12	13
6.5	42.25	42.38	42.51	42.64	42.77	42.90	43.03	43.16	43.30	43.43	13
6.6	43.56	43.69	43.82	43.96	44.09	44.22	44.36	44.49	44.62	44.76	13
6.7	44.89	45.02	45.16	45.29	45.43	45.56	45.70	45.83	45.97	46.10	14
6.8	46.24	46.38	46.51	46.65	46.79	46.92	47.06	47.20	47.33	47.47	14
6.9	47.61	47.75	47.89	48.02	48.16	48.30	48.44	48.58	48.72	48.86	14
7.0	49.00	49.14	49.28	49.42	49.56	49.70	49.84	49.98	50.13	50.27	14
7.1	50.41	50.55	50.69	50.84	50.98	51.12	51.27	51.41	51.55	51.70	14
7.2	51.84	51.98	52.13	52.27	52.42	52.56	52.71	52.85	53.00	53.14	15
7.3	53.29	53.44	53.58	53.73	53.88	54.02	54.17	54.32	54.46	54.61	15
7.4	54.76	54.91	55.06	55.20	55.35	55.50	55.65	55.80	55.95	56.10	15
7.5	56.25	56.40	56.55	56.70	56.85	57.00	57.15	57.30	57.46	57.61	15
7.6	57.76	57.91	58.06	58.22	58.37	58.52	58.68	58.83	58.98	59.14	15
7.7	59.29	59.44	59.60	59.75	59.91	60.06	60.22	60.37	60.53	60.68	16
7.8	60.84	61.00	61.15	61.31	61.47	61.62	61.78	61.94	62.09	62.25	16
7.9	62.41	62.57	62.73	62.88	63.04	63.20	63.36	63.52	63.68	63.84	16
8.0	64.00	64.16	64.32	64.48	64.64	64.80	64.96	65.12	65.29	65.45	16
8.1	65.61	65.77	65.93	66.10	66.26	66.42	66.59	66 75	66.91	67.08	16
8.2	67.24	67.40	67.57	67.73	67.90	68.06	68.23	68.39	68.56	68.72	17
8.3	68.89	69.06	69.22	69.39	69.56	69.72	69.89	70.06	70.22	70.39	17
8.4	70.56	70.73	70.90	71.06	71.23	71.40	71.57	71.74	71.91	72.08	17
8.5	72.25	72.42	72.59	72.76	72.93	73.10	73.27	73.44	73.62	73.79	17
8.6	73.96	74.13	74.30	74.48	74.65	74.82	75.00	75.17	75.34	75.52	17
8.7	75.69	75.86	76.04	76.21	76.39	76.56	76.74	76.91	77.09	77.26	18
8.8	77.44	77.62	77.79	77.97	78.15	78.32	78.50	78.68	78.85	79.03	18
8.9	79.21	79.39	79.57	79.74	79.92	80.10	80.28	80.46	80.64	80.82	18
9.0	81.00	81.18	81.36	81.54	81.72	81.90	82.08	82.26	82.45	82.63	18
9.1	82.81	82.99	83.17	83.36	83.54	83.72	83.91	84.09	84.27	84.46	18
9.2	84.64	84.82	85.01	85.19	85.38	85.56	85.75	85.93	86.12	86.30	19
9.3	86.49	86.68	86.86	87.05	87.24	87.42	87.61	87.80	87.98	88.17	19
9.4	88.36	88.55	88.74	88.92	89.11	89.30	89.49	89.68	89.87	90.06	19
9.5	90.25	90.44	90.63	90.82	91.01	91.20	91.39	91.58	91.78	91.97	19
9.6	92.16	92.35	92.54	92.74	92.93	93.12	93.32	93.51	93.70	93.90	19
9.7	94.09	94.28	94.48	94.67	94.87	95.06	95.26	95.45	95.65	95.84	20
9.8	96.04	96.24	96.43	96.63	96.83	97.02	97.22	97.42	97.61	97.81	20
9.9	98.01	98.21	98.41	98.60	98.80	99.00	99.20	99.40	99.60	99.80	20
n	0	1	2	3	4	5	6	7	8	9	Diff.

An Introduction to the Approximation of Functions, Theodore J. Rivlin. Unabridged republication of the 1969 edition. 160pp. 49554-X

An Essay on the Foundations of Geometry, Bertrand Russell. Unabridged republication of the 1897 edition. 224pp. 49555-8

Elements of Number Theory, I. M. Vinogradov. Unabridged republication of the first edition, 1954. 240pp. 49530-2

Asymptotic Expansions for Ordinary Differential Equations, Wolfgang Wasow. Unabridged republication of the 1976 corrected, slightly enlarged reprint of the original 1965 edition. 384pp. 49518-3

Physics

Semiconductor Statistics, J. S. Blakemore. Unabridged, corrected, and slightly enlarged republication of the 1962 edition. 141 illustrations. xviii+318pp. 49502-7

Wave Propagation in Periodic Structures, L. Brillouin. Unabridged republication of the 1946 edition. 131 illustrations. 272pp. 49556-6

The Conceptual Foundations of the Statistical Approach in Mechanics, Paul and Tatiana Ehrenfest. Unabridged republication of the 1959 edition. 128pp. 49504-3

The Analytical Theory of Heat, Joseph Fourier. Unabridged republication of the 1878 edition. 20 figures. 496pp. 49531-0

States of Matter, David L. Goodstein. Unabridged republication of the 1975 edition. 154 figures. 4 tables. 512pp. 49506-X

The Principles of Mechanics, Heinrich Hertz. Unabridged republication of the 1900 edition. 320pp. 49557-4

Thermodynamics of Small Systems, Terrell L. Hill. Unabridged and corrected republication in one volume of the two-volume edition published in 1963–1964. 32 illustrations. 408pp. 6½ x 9¼. 49509-4

Theoretical Physics, A. S. Kompaneyets. Unabridged republication of the 1961 edition. 56 figures. 592pp. 49532-9

Quantum Mechanics, H. A. Kramers. Unabridged republication of the 1957 edition. 14 figures. 512pp. 49533-7

The Theory of Electrons, H. A. Lorentz. Unabridged reproduction of the 1915 edition. 9 figures. 352pp. 49558-2

The Principles of Physical Optics, Ernst Mach. Unabridged republication of the 1926 edition. 279 figures. 10 portraits. 336pp. 49559-0

The Scientific Papers of James Clerk Maxwell, James Clerk Maxwell. Unabridged republication of the 1890 edition. 197 figures. 39 tables. Total of 1,456pp. *Volume I* (640pp.) 49560-4; *Volume II* (816pp.) 49561-2

Vectors and Tensors in Crystallography, Donald E. Sands. Unabridged and corrected republication of the 1982 edition. xviii+228pp. 49516-7

Principles of Mechanics and Dynamics, Sir William (Lord Kelvin) Thompson and Peter Guthrie Tait. Unabridged republication of the 1912 edition. 168 diagrams. Total of 1,088pp. *Volume I* (528pp.) 49562-0; *Volume II* (560pp.) 49563-9

Treatise on Irreversible and Statistical Thermophysics: An Introduction to Nonclassical Thermodynamics, Wolfgang Yourgrau, Alwyn van der Merwe, and Gough Raw. Unabridged, corrected republication of the 1966 edition. xx+268pp. 49519-1

Engineering

Principles of Aeroelasticity, Raymond L. Bisplinghoff and Holt Ashley. Unabridged, corrected republication of the original 1962 edition. xi+527pp. 49500-0

Statics of Deformable Solids, Raymond L. Bisplinghoff, James W. Mar, and Theodore H. H. Pian. Unabridged and corrected Dover republication of the edition published in 1965. 376 illustrations. xii+322pp. 6½ x 9¼. 49501-9